Optimizing Molecular Methods to Detect Human Caliciviruses in Environmental Samples

Awwa Research Foundation

About the Awwa Research Foundation

The Awwa Research Foundation (AwwaRF) is a member-supported, international, nonprofit organization that sponsors research to enable water utilities, public health agencies, and other professionals to provide safe and affordable drinking water to consumers.

The Foundation's mission is to advance the science of water to improve the quality of life. To achieve this mission, the Foundation sponsors studies on all aspects of drinking water, including supply and resources, treatment, monitoring and analysis, distribution, management, and health effects. Funding for research is provided primarily by subscription payments from approximately 1,000 utilities, consulting firms, and manufacturers in North America and abroad. Additional funding comes from collaborative partnerships with other national and international organizations, allowing for resources to be leveraged, expertise to be shared, and broad-based knowledge to be developed and disseminated. Government funding serves as a third source of research dollars.

From its headquarters in Denver, Colorado, the Foundation's staff directs and supports the efforts of more than 800 volunteers who serve on the board of trustees and various committees. These volunteers represent many facets of the water industry, and contribute their expertise to select and monitor research studies that benefit the entire drinking water community.

The results of research are disseminated through a number of channels, including reports, the Web site, conferences, and periodicals.

For subscribers, the Foundation serves as a cooperative program in which water suppliers unite to pool their resources. By applying Foundation research findings, these water suppliers can save substantial costs and stay on the leading edge of drinking water science and technology. Since its inception, AwwaRF has supplied the water community with more than $300 million in applied research.

More information about the Foundation and how to become a subscriber is available on the Web at **www.awwarf.org.**

Optimizing Molecular Methods to Detect Human Caliciviruses in Environmental Samples

Prepared by:
Jan Vinjé
University of North Carolina at Chapel Hill
Chapel Hill, NC, 27599

Sponsored by:
Awwa Research Foundation
6666 West Quincy Avenue, Denver, CO 80235-3098
and
U.S. Environmental Protection Agency
Washington D.C.

Published by:

Awwa Research Foundation

American Water Works Association

IWA Publishing

DISCLAIMER

This study was jointly funded by the Awwa Research Foundation (AwwaRF) and the U.S. Environmental Protection Agency (USEPA) under Cooperative Agreement No. R82940901. AwwaRF and USEPA assume no responsibility for the content of the research study reported in this publication or for the opinions or statements of facts expressed in the report. The mention of trade names for commercial products does not represent or imply the approval or endorsement of AwwaRF or USEPA. This report is presented solely for information purposes.

CONTENTS

TABLES

FIGURES

FOREWORD

The Awwa Research Foundation is a nonprofit corporation that is dedicated to the implementation of a research effort to help utilities respond to regulatory requirements and traditional high-priority concerns of the industry. The research agenda is developed through a process of consultation with subscribers and drinking water professionals. Under the umbrella of a Strategic Research Plan, the Research Advisory Council prioritizes the suggested projects based upon current and future needs, applicability, and past work; the recommendations are forwarded to the Board of Trustees for final selection. The foundation also sponsors research projects through the unsolicited proposal process; the Collaborative Research, Research Applications, and Tailored Collaboration programs; and various joint research efforts with organizations such as the U.S. Environmental Protection Agency, the U.S. Bureau of Reclamation, and the Association of California Water Agencies.

This publication is a result of one of these sponsored studies, and it is hoped that its findings will be applied in communities throughout the world. The following report serves not only as a means of communicating the results of the water industry's centralized research program but also as a tool to enlist the further support of the nonmember utilities and individuals.

Projects are managed closely from their inception to the final report by the foundation's staff and large cadre of volunteers who willingly contribute their time and expertise. The foundation serves a planning and management function and awards contracts to other institutions such as water utilities, universities, and engineering films. The funding for this research effort comes primarily from the Subscription Program, through which water utilities subscribe to the research program and make an annual payment proportionate to the volume of water they deliver and consultants and manufacturers subscribe based on their annual billings. The program offers a cost-effective and fair method for funding research in the public interest.

A broad spectrum of water supply issues is addressed by the foundation's research agenda: resources, treatment and operations, distribution and storage, water quality and analysis, toxicology, economics, and management. The ultimate purpose of the coordinated effort is to assist water suppliers to provide the highest possible quality of water economically and reliably. The true benefits are realized when the results are implemented at the utility level. The foundation's trustees are pleased to offer this publication as a contribution toward that end.

David E. Rager
Chair, Board of Trustees
Awwa Research Foundation

Robert C. Renner, P.E.
Executive Director
Awwa Research Foundation

ACKNOWLEDGMENTS

We would like to thank the Awwa Research Foundation for funding this project. Part of this project was also funded by a Grant from the European Union. We would like to thank the Project Advisory Committee for helpful suggestions and constructive criticism.

The laboratory work described in this report was conducted by Jennifer Cannon, Andrea Maloney as part of their Master's of Public Health thesis and by Dr. Jothikumar Narayanan. We thank Drs. Steve Monroe, Joanne Bartkus, Kjell-Olof Hedlund, Kellogg Schwab, Chuck Gerba, Christine Moe, Jason Jiang and the NC State Health Laboratory, Raleigh NC for providing us with norovirus positive stool specimens. Many thanks also to Columbus Water Works, Columbus, GA, Cobb County-Marietta Water Authority, Acworth, GA and waste water treatment plants City of Atlanta, GA and Cobb County Water System, Marietta, GA for providing us with water samples.

EXECUTIVE SUMMARY

The most common cause of acute gastroenteritis in humans are the human caliciviruses of which sapoviruses infect mainly infants and young children and noroviruses infect people of all ages. Noroviruses are the most important cause of outbreaks of acute nonbacterial gastroenteritis and are associated with 50% of foodborne outbreaks. Exposure to contaminated water has been reported as an important transmission route and several waterborne norovirus outbreaks have been documented. Although noroviruses may be present in the environment in very low quantities, they are a significant health risk as the minimum infectious dose for norovirus infection is in the range 10-100 virus particles. Most published molecular methods are based on the detection of noroviruses in stool samples and limited data are available on the sensitivity of detection of the particular primer sets, which may differ significantly for viruses belonging to different genetic clusters of noroviruses.

In this project, we proposed to develop and evaluate the use of ligand binding assays based on H-type carbohydrates, which recently have been shown to bind to norovirus virus-like particles as a proof of concept for the recovery of noroviruses from environmental water samples. We compared and evaluated this method along with an electropositive 1MDS filter-elution method using recently improved amino acid elution media. We explored different strategies for the development of a real-time RT-PCR assay for the detection of noroviruses genogroup I and genogroup II viruses and validated assays using a comprehensive panel of genetically different norovirus strains. Further, we systematically evaluated and compared the performance of conventional reverse transcription-polymerase chain reaction (RT-PCR) assays using a variety of existing primer pairs targeting different regions of the norovirus genome. Successful candidate methods both conventional as well as real-time were validated on different environmental sample matrices (drinking water, waste water, shellfish).

RESEARCH OBJECTIVES

Widespread use of molecular methods for the detection of noroviruses (NoVs) in water has been hampered by the unavailability of sensitive and robust methods that are able to detect this group of genetically diverse viruses at low copy numbers. Although at present several broadly reactive RT-PCR assays targeting different regions of the viral genome have been described, these are primarily optimized for the detection of NoVs in fecal specimens with a relatively high viral load. This project evaluated currently used primer pairs targeting different regions of the viral genome using conventional RT-PCR methodologies and explored the same regions for candidate real-time RT-PCR approaches. The overall goal of this project was to select an optimized method that detects the majority of the recognized NoV genogroups and genetic clusters that is applicable to a wide range of environmental samples. The specific objectives of this project include:

- Developing and evaluating a carbohydrate ligand-binding assay for the purification of noroviruses from RT-PCR interfering substances as a possible alternative or addition to RNA extraction methods currently used for different water matrices.
- Developing a real-time RT-PCR assay for noroviruses.
- Comparing and evaluating previously published and recently developed primer pairs targeting different regions of the norovirus genome for their broadness and sensitivity to the more than 20 different norovirus clusters in a conventional RT-PCR assay.

- Using the developed methods for concentration, purification and molecular detection to examine environmental samples (shellfish, source and finished water). The source and finished waters include fecally contaminated water with human wastes (sewage) and source and finished waters from drinking water treatment plants.

APPROACH

1. Develop and evaluate a carbohydrate ligand-binding assay for norovirus recovery and purification

   Recently, several groups have discovered that recombinant Norwalk virus-like particles (NV VLPs) bind to histo-blood group antigens present on gastroduodenal epithelial cells and saliva and they agglutinate most human red blood cells (Marionneau et al., 2002, Hutson et al., 2002a, Lindesmith et al., 2003). Further studies showed that rNV VLPs bind to synthetic H and Lewis carbohydrate antigens (Hutson et al., 2002). Therefore, synthetic carbohydrates that have a high affinity for rNV VLPs can be bound to silica beads or magnetic beads and used for (immunomagnetic) separation similar as has been described for beads coated with antibodies (Myrmel et al., 2000, Jothikumar et al., 1998, Schwab et al., 1996). Without an available cell culture system for noroviruses, capture of complete virions by antibodies or carbohydrate receptors would be the best surrogate for detecting complete and possibly infectious virus particles as has been described for rabbit hemorrhagic disease virus (Ruvoen-Clouet et al., 2000). To determine if synthetic H and Lewis carbohydrate antigens are able to bind different clusters of noroviruses, we have tested a panel of GI.1 positive stool samples by absorption with immobilized synthetic oligosaccharides onto avidine coated magnetic beads (M-270 Amine Dynabead, Dynal Biotech). Immunomagnetic separation (IMS) was then employed to evaluate the performance of this method for the recovery of low-copy number of NoVs from environmental samples.

2. Develop a real-time RT-PCR assay for GI and GII noroviruses

   Based on our sequence database of different regions of the NoV genome and on recent published literature (Kageyama et al., 2003), we identified the ORF1-ORF2 junction region to be the most conserved to develop a TaqMan realtime RT-PCR assay. When concentrating viruses from large volumes of water (100-L or greater), organic (humic acids, fulvic and tannic acid, polysaccharides, proteins, glycoproteins) and inorganic compounds (cations) are co-concentrated and present in the final concentrate. For environmental samples we used a recently developed lysis buffer for extraction of RNA from environmental samples, which is compatible with a variety of the commercially available silica gel based columns (Roche, Qiagen and Omega Bio-Tek) and we have found it compatible with fluorogenic probe-based RT-PCR assays for noroviruses (LightCycler and Smart Cycler).

3. Comparison of Conventional and TaqMan® Realtime RT-PCR assays for the detection of Human Noroviruses in Environmental Water Samples

   Molecular detection of NoVs has found widespread use since these viruses have to date been impossible to grow in cell culture. After the successful cloning and sequencing of the complete Norwalk virus genome, first generation primers met little success when applied to outbreak specimens that were positive for these viruses by electron microscopy. Since then, several groups have targeted conserved amino acid motifs in the RNA-polymerase gene (region A), and later also other regions of the genome (region B, C and D). Noroviruses can be grouped in 5 genogroups that can be further delineated into at least 29 genetic clusters or genotypes based on grouping by genetic relatedness in the complete capsid gene (Zheng et al., 2006). Although many different primer sets have been successfully used for detecting noroviruses in outbreak studies (Ando et al., 1995, Vinjé et al., 1997, Maguire et al., 1999, Fankhauser et al., 2002), few studies have been reported that describe the quantity of virus genome that minimally must be present for a primer pair to detect a virus (Le Guyader et al., 1996). We systematically compared different published oligonucleotide primer sets for NoVs and compared the hemi-nested assays with our TaqMan® Realtime RT-PCR assays.

CONCLUSIONS

In this study we developed and evaluated novel methods for the concentration, detection and genotyping of NoVs from complex environmental matrices. Specific conclusions of the project are:

1. As a proof-of-concept, we developed a novel assay based on H-type 1 histo-bloodgroup carbohydrate magnetic beads (HGBA-beads) which had a detection limit of 20-200 RT-PCR detectable units of GI.1 norovirus.
2. GI.1 NoV particles bound to the HGBA-beads were resistant to RNase treatment indicating the protein capsid of the bound viruses is intact and thus potentially infectious.
3. This study provides the framework for implementing CMS for concentration and recovery of low copy numbers of NoV from environmental waters for detection by molecular methods. We hypothesize that binding of human NoV to HBGAs followed by detection by RT-PCR may be a surrogate for detecting infectious viruses.
4. In the present study, we have developed a rapid (less than 90 min.) and broadly reactive one-step TaqMan® RT-PCR assays for the detection of GI and GII NoV in fecal samples as well as shellfish samples. The specificity and sensitivity of all steps of the assays were systematically evaluated and in its final format, the monoplex assays were validated using RNA extracted from a panel of 84 stool specimens including NoV strains representing 19 (7 GI , 11 GII, and 1 GIV) different genotypes. The assays were further validated on 38 shellfish cDNA extracts previously tested by nested PCR.
5. Comparison with a recently published real-time assay showed that our one-step TaqMan® RT-PCR assays had a significantly higher sensitivity, and was at least as sensitive as nested PCR. For stool specimens, a duplex TaqMan® RT-PCR assay using GI and GII TaqMan® probes labeled with FAM and JOE, respectively, performed equally well compared to individual genogroup-specific monoplex assays. All other tested enteric viruses tested negative and no cross-reaction between genogroups was observed.
6. In this study, we compared the detection limit of 13 (7 GI, 6 GII) different norovirus oligonucleotide primer combinations targeting the 5'-end of ORF2 (region C) by conventional RT-PCR followed by hemi-nested PCR. A panel of 24 norovirus strains including four different GI genotypes (GI.1, GI.2, GI.3, GI.4) and seven different GII genotypes (GII.1 through GII.7) was used to evaluate the different primer combinations of which COG1F/G1SKR and G1SKF/G1SKR for GI and COG2F/G2SKR and

G2SKF/G2SKR for GII NoV showed the lowest detection limit across the range of genotypes tested.

7. We compared these hemi-nested PCR assays with a recently developed TaqMan®-based real-time RT-PCR assay (see previous chapter) on naturally contaminated water samples collected from Sept 2003 –Sept 2004. None of the drinking water samples and 63% of the waste water samples tested positive for NoVs. While hemi-nested PCR is as sensitive as real-time PCR detection and is capable to link clinical and environmental strains by sequence analysis of the PCR products, TaqMan® real-time RT-PCR is a superior method for rapid and simple monitoring of environmental waters for NoVs.

FUTURE RESEARCH

Additional research is needed on identifying a cocktail of HBGAs that are able to bind to the majority of the epidemiological important human norovirus strains including GII.4. Further field testing of the methods developed in this study is recommended i.e., testing water implicated in clusters of norovirus disease.

CHAPTER 1
INTRODUCTION

Infections with noroviruses (NoVs), previously called "Norwalk-like viruses", are the most common cause of outbreaks of acute non-bacterial gastroenteritis in humans worldwide. Based on recent studies, NoVs account for 94% of the reported outbreaks that have been examined over a 4.5-year period in the US (Fankhauser et al., 2002), 87% of all outbreaks reported in The Netherlands (Vinjé et al., 1997) and similar high percentages in other outbreak studies (Hedlund et al., 2000; Maguire et al., 1999; Nakata et al., 2000; Gonin et al., 2000). Recent estimates suggest that of the 38.6 million annual cases of gastroenteritis in the US, 80% are the result of viral infections of which the majority is attributed to NoVs (Mead et al., 1999). The use of molecular methods including reverse transcriptase-polymerase chain reaction (RT-PCR) and antigen detecting enzyme immunoassays have changed our understanding about the burden and epidemiology of NoV and sapoviruses, previously called Sapporo-like viruses. NoVs not only account for the majority of outbreaks but are also as common as rotaviruses among children less than 2 years of age (Pang et al., 1999). Conversely, sapoviruses have been primarily associated with pediatric gastroenteritis in disease-endemic settings and are rarely associated with outbreaks.

The primary route of transmission of NoVs is person-to-person but transmission via contaminated, common-source vehicles, especially food and water, has increasingly been documented. Common-sources are of particular importance as potentially large numbers of people are involved and there are risks for multistate or international outbreaks (Lawson et al., 1991, Beller et al., 1997, Kukkula et al., 1999, Schvoerer et al., 1999, Hafliger et al., 2000, Boccia et al., 2002, Ponka et al., 1999, Anderson et al., 2001, Koopmans et al., 2002). NoVs have also been implicated in many outbreaks associated with the consumption of shellfish (Shieh et al., 2000). Although NoV associated outbreaks can be found throughout the year, a cold weather peak was demonstrated in the majority of studies conducted in 8 different countries including the US (Mounts et al., 2000).

The cloning and sequencing of Norwalk, Southampton and Manchester virus (Jiang et al., 1990, Lambden et al., 1993, Liu et al., 1995) has allowed the classification of these viruses within two separate genera of the family *Caliciviridae*. Norwalk virus and related viruses were classified within the genus *Norovirus* and Manchester and Sapporo viruses within the genus *Sapovirus* (Green et al., 2007). Based on sequence information obtained from the RNA polymerase gene or the major capsid protein gene (VP1), NoVs can be further subdivided into 5 separate genetic groups or genogroups of which viruses of genogroup III have so far only been found in cattle (Ando et al., 2000) and genogroup V only in mice (Karst and Wobus et al, 2003). Within genogroups strains can be further delineated into at least 30 genetic clusters or genotypes based on grouping by genetic relatedness in the complete capsid gene.

Despite the lack of a cell culture system for NoV, availability of nucleic methods and their use to develop nucleotide sequence data has facilitated the development of sensitive molecular diagnostics for virus detection in stool samples. First generation RT-PCR assays that used primers based on the prototype Norwalk virus gave poor performance because the genetic diversity of NoV strains was much greater than initially anticipated (De Leon et al., 1992, Jiang et al., 1992, Moe et al., 1994). Therefore the primers selected were further optimized. Using sequence information of an increasing number of NoV strains, several research groups successfully developed RT-PCR assays based on improved primers targeting a conserved region of ORF1 coding for the viral RNA polymerase (Ando et al., 1995, Green J et al., 1995, Green S et al., 1995, Jiang et al., 1999, Le Guyader et al., 1996, Vinjé and Koopmans 1996). Subsequently, different primer sets targeting region A have been used successfully in epidemiological studies for the diagnosis of NV in fecal specimens from both outbreaks and sporadic cases (Ando et al., 1995, de Wit et al., 2001, Green J et al., 1995, Maguire et al., 1999, Vinjé and Koopmans 1996 1997). In addition, primers directed to other regions of the NV genome have been developed including relatively conserved regions at the 3'-end of ORF1 (Fankhauser et al., 2002) and in the capsid protein gene (region C and D). The capsid regions were selected as regions better reflecting presumed antigenicity of different NV types (Green S et al., 1995, Noel et al., 1997, Vinjé et al., 2000a, Wang et al., 1994).

Although successful for the generic detection of NoV in stool samples, different primer pairs can have significant different detection limits (up to 10^4) for certain genotypes (Vinje et al., 2003). Thus, the choice of primer pair could greatly influence the limit of detection for each individual genotype, which is anticipated to be crucial when detecting low copy numbers of NoV in environmental samples.

Nevertheless, several groups have successfully detected NoV genes in implicated water samples often using nested PCR formats or developing strain specific primer pairs to lower the limit of detection (Hafliger et al., 1997; Beller et al., 1997; Boccia et al., 2002).

Real-time RT-PCR is an attractive alternative to conventional PCR due to its greater rapidity, sensitivity, and reproducibility (Mackay et al., 2002). Typically, real-time detection reduces hands-on time and, as post-amplification manipulation is not required, decreases risk of contamination from amplified product. The assay is based on an increase in fluorescence that is proportional to the amount of amplicon production during each PCR cycle (i.e., in real time), and unless SYBR Green is used, is not influenced by non-specific amplification. SYBR Green chemistry however, does not require the design or synthesis of probes, is the least expensive assay, and has been used for certain applications like routine monitoring of viruses with limited genetic variability (e.g., hepatitis A virus). Greater specificity and quantitative results can be obtained using sequence-specific internal oligonucleotide probes (Taqman, molecular beacons, scorpion probes etc.). In addition, the use of real-time PCR has provided insights into the role that some compounds have on PCR inhibition from a diverse range of sample types, including stool samples (Belanger et al., 2002).

Successful methods to concentrate and purify NoV from large volumes of water (~100-L) include adsorption of viruses to positively charged disk or cartridge filters and elution with beef-extract at pH 9.0 (Beller et al., 1997; Sobsey and Glass 1980; Sobsey and Hickey 1985) or with 0.5 M of Lysine (Hill et al., 2002). Because beef extract is chemically ill defined, different amino acids have been tested as successful alternatives (Beller et al., 1997; Hill et al., 2002). The viruses can be further concentrated and purified by PEG precipitation and chloroform extraction as alternative to acid precipitation of beef extract (Jiang et al., 2001; Shieh et al., 1997). Most methods developed for the recovery of viral RNA involve multistep extraction and elution. Lysis of the virus by guanidinium thiocyanate (GuSCN) and binding to silica-based membranes has become a standard method in most clinical laboratories. For environmental samples with high turbidity, DNAse pretreatment of the samples or the Trizol extraction method that includes both GuSCN for lysis of the virus and phenol/chloroform extraction of viral RNA might be suitable for obtaining a clean viral RNA extract without (RT)-PCR inhibitors. The use of an internal RNA control is the best way to effectively determine whether inhibitors are present in a particular sample (Atmar et al., 1995; Schwab et al., 1997). However, previously described internal controls are too genotype-specific to be used efficiently in RT-PCR amplification of genetically diverse NoVs.

Recently, several groups have discovered that recombinant Norwalk virus (rNoV) virus-like particles bind to histo-blood group antigens present on gastroduodenal epithelial cells and saliva and they agglutinate most human red blood cells (Marionneau et al., 2002; Hutson et al., 2002; Lindesmith et al., 2003). Further studies showed that rNoV VLPs bind to synthetic H and Lewis carbohydrate antigens (Hutson et al., 2002b). Therefore, synthetic carbohydrates that bind to rNoV VLPs can be (covalently) bound to silica beads or magnetic beads and used for (immunomagnetic) separation similar as has been described for beads coated with antibodies (Myrmel et al., 2000; Jothikumar et al., 1998; Schwab et al., 1996). Without an available cell culture system for noroviruses, capture of complete virions by antibodies or carbohydrate ligands would be the best surrogate for detecting complete and possibly infectious virus particles as has been described for rabbit hemorrhagic disease virus (Ruvoen-Clouet et al., 2000).

CHAPTER 2
NOVEL CARBOHYDRATE LIGAND-BINDING ASSAY FOR THE RECOVERY AND PURIFICATION OF NOROVIRUS FROM WATER CONCENTRATES

INTRODUCTION

Now recognized as the leading cause of outbreaks as well as sporadic cases of acute gastroenteritis worldwide, human noroviruses (NoVs) contribute to an estimated 67% of all foodborne illnesses in the United States (Mead et al., 1999). While person to person and foodborne transmissions are common, waterborne spread has gained attention since its induction onto the EPAs drinking water contaminant list. Water becomes contaminated after contact with stool, vomit (including aerosols), soiled hands, soiled clothing or contaminated surfaces or infiltration of untreated or inadequately treated water or sewage. Exposure to contaminated drinking water, recreational waters or ice has led to several large waterborne outbreaks of NoV (Hoebe et al., 2004; Khan, et al., 1994; Maunula et al., 2004) and contamination of irrigation waters, shellfish harvesting waters and water used for food washing has lead to foodborne outbreaks (Berg et al., 2000; Croci et al., 2007; McDonnell et al., 1997; Shieh et al., 2003).

Because no routine cell culture assay is available for human noroviruses, norovirus RNA is detected using molecular methods such as RT-PCR and real time RT-PCR. Although environmental waters may contain low concentrations of norovirus, exposure to even a small number of NoVs is a public health risk as the infectious dose is very low (10-100 virus particles) (Moe et al., 1999). Therefore, to test potentially contaminated water, large sample volumes (1 L to 1000 L) must be concentrated into microliter volumes. Environmental are complex matrices that often contain elements that interfere with molecular techniques when co-concentrated with the virus. Methods are needed to purify the these waters prior to application of molecular detection techniques in order to avoid false negative results.

Recently, histo-blood group antigens (HBGAs) have been identified as markers of genetic susceptibility and resistance for norovirus infection as symptomatic Norwalk virus infection of human volunteers is dependent on cells displaying the appropriate HBGA receptor(s) on cells of the gut mucosa (Hutson et al., 2003; Lindesmith et al., 2003). Individuals that do not display certain HBGAs on their intestinal mucosa or its secretions have been shown to be resistant to symptomatic infection. Several, but not all, NoVs and their respective virus-like particles (VLPs) bind to synthetic HBGAs and binding can be blocked by hyperimmune sera or saliva from secretors (Harrington et al., 2002; Harrington et al., 2004; Huang et al., 2003; Hutson et al., 2002; Lindesmith et al., 2003). Furthermore, dimerization of the NoV P2 domains of adjacent VP1 proteins form a specific receptor binding pocket for HBGAs that can be disrupted by protein mutagenesis of amino acids in this region (Tan et al., 2003; Tan et al., 2004) and a crystallographic structure of a genogroup II norovirus binding to HBGA has been solved (Caul et al., 2007)).

In this proof-of-concept study we developed and evaluated a HBGA binding assay for detecting of GI.1 NoVs in environmental waters using a magnetic separation technique (CMS). Bound particles were RNase resistant and therefore we hypothesize that CMS may be a surrogate for detecting infectious viruses. Overall the results of this study provide a framework for implementing CMS for concentration and recovery of low copy numbers of NoV from environmental waters.

MATERIALS AND METHODS

Viruses and environmental water samples

We used a panel (n = 5) of GI.1 norovirus positive stool samples obtained from human volunteers studies which were kindly provided by Drs Christine Moe and Robert Atmar. Virus titers were estimated based on conventional (De Leon et al., 1992) and realtime RT-PCR (Jothikumar et al., 2005). A stool norovirus positive stool sample genotyped as GII.4 (Vinje et al., 2004) was used as a control in the RNase protection assay. Group A rotavirus strain WA (serotype 1) and human astrovirus serotype 1 were included to evaluate the specificity of the assay. Concentrates of surface water (influent drinking water) (n=4) and influent (n=2) and effluent (n=4) wastewater samples were collected from drinking and waste water facilities in Columbus, GA and Atlanta, GA, respectively and stored at 4°C before processing.

Carbohydrate Magnetic Separation (CMS)

Streptavidin-coated magnetic beads (25 µl) (Dynal Biotech, Olso, Norway) were washed 3 times in PBS and pre-blocked with blocking buffer [5% fetal bovine serum (FBS) or SuperBlock (Pierce Biotech., Rockford, IL)] for 1 hr before adding 50 µl of synthetic biotinylated H type 1 histo-blood group carbohydrates (1 mg/ml) (Glycotech, Rockville, MD) in 950 µl of blocking buffer [5% fetal bovine serum (FBS) or SuperBlock (Pierce Biotech., Rockford, IL)] for 1.5 hr at room temperature. Beads were subsequently washed 3 times with 0.05% PBS-Tween 20. A magnet (Dynal Biotech, Oslo, Norway) accommodating 1.7 ml microcentrifuge tubes was used during washing steps. Beads were then blocked for 1 hr at room temperature followed by overnight incubation at 4°C in 1 ml blocking buffer [5% fetal bovine serum (FBS) or SuperBlock (Pierce Biotech., Rockford, IL)] supplemented with 0.25% Tween-20. After 3 washes with 0.05% PBS Tween-20, the beads were suspended in 1 ml of blocking buffer, environmental water concentrate, or PBS. Two µl of ten-fold serial dilutions of 10% clarified stool suspension were added and the suspensions were incubated for 4 hr at room temperature on an end-over-end rotator (Dynal Biotech, Oslo, Norway). The beads were then washed 5 times with 0.05% PBS Tween-20 followed by 3 times with PBS and then suspended in 50 µl PBS. The suspensed beads were then 10-fold serially diluted in PBS and RNA was extracted by heat release (5 min. at 99°C). Beads were pelleted by a 25 sec spin at 14,000 x g and 2.5 µl or 1 µl of supernatant was analyzed by conventional or TaqMan realtime RT-PCR, respectively. Serial dilutions of viral RNA (Norwalk) were used as positive control and RNase-free water as negative controls.

Conventional and TaqMan realtime Reverse-Transcription Polymerase chain reaction

Conventional RT-PCR was performed using a Qiagen One-Step RT-PCR kit (Qiagen, Valencia, CA) including the Q solution and using Norwalk virus specific primers (pol5' and pol 3'; 15 µM each) (De Leon et al., 1992) targeting the polymerase region of the genome. Cycling conditions included a 42°C RT step for 1 hr, 15 min denaturation at 95°C followed by 35 cycles of 94°C, 50°C, and 72°C for 30 sec each and ending with a 7 min 72°C extension and 4°C hold. TaqMan realtime RT-PCR was performed using an AgPath ID One-Step RT-PCR kit (Ambion, Austin, TX) without the detection enhancer solution on the ABI 7900 (Applied Biosystems, Foster City, CA) using GI primers (CogI F/R (10 µM each) and FAM labeled probes (Ring 1a and ring 1 b (1.5 µM each) targeting the conserved ORF1 - ORF2 junction region of the genome (Kageyama et al., 2003; Trujillo et al., 2006). Cycling conditions were as follows: reverse transcription at 45°C for 10 min, activation of Taq polymerase at 95°C for 10 min, followed by 55 cycles of 95°C for 15 sec and 65°C for 1 min. Data were collected during the 65°C extension period for each cycle.

Detection limit of CMS

Replicate 10-fold serial dilutions of viral RNA extracted by a Qiagen viral RNA stool kit were used to estimate the RNA titer in each positive stool sample. RT-PCR-detectable units (PDU) were calculated based on the highest dilution of RNA giving a positive band on an ethidium bromide stained 2% agarose gel or the highest dilution detectable by real-time RT-PCR.

Stool suspensions were 10-fold serially diluted in RNase-free water. To determine the detection limit of CMS, 2,000, 200, 20 or 2 PDU in SuperBlock blocking buffer were added to carbohydrate-bound beads and analyzed as described above. Following heat release, undiluted and 10 times diluted beads in RNase-free water were analyzed by conventional RT-PCR for end-point detection and by real-time RT-PCR for quantitation.

Specificity of CMS

Two sets of CMS reaction tubes were prepared containing 2,000, 200, and 20 PDU of Norwalk virus (8FIIb; 23-3) in 1 ml of SuperBlock. To one set of tubes, 10^6 infectious rotavirus group A serotype 1 was added and 5 x 10^5 human astrovirus (serotype 1) was added to the other set. Inhibition of Norwalk virus binding to the carbohydrate beads was measured by comparison with controls with Norwalk virus only.

Concentration of environmental waters

Surface waters

Four liters of surface water (influent drinking water) were pH adjusted to 7.0-7.2 followed by polyethylene glycol (PEG) precipitation (8% PEG 8000, 0.3 M NaCl) at 4°C overnight with constant stirring. PEG-precipitated virus was collected by centrifugation at 6,735 x g for 1 hr at 4°C. Samples were suspended in PBS pH 7.2 and extracted with 50% (v/v) chloroform, centrifuged at 2000 x g for 10 min at 4°C. The supernatants (10-20 ml) were stored at -70°C.

Waste waters

Eight hundred fifty milliliters of influent (raw) and effluent wastewater was centrifuged at 6,735 x g for 1 hr at 4°C. Retaining the supernatant, the pellet was chloroform extracted (50% (v/v)) and centrifuged at 2,000 x g for 10 min at 4°C. The aqueous phase was combined with the supernatant from the first centrifugation and the samples were processed by PEG precipitation as above. The supernatants (5-10 ml) stored at -70°C.

Nuclease protection assay

To determine if the Norwalk virus particles bound to H-type 1 coated beads were completely intact and potentially infectious, 20,000 or 4,000 PDU of NoV-1 were treated with 10 ng of ribonuclease A (Invitrogen, Carlsbad, CA) in 0.1 M TE buffer and incubated for 30 min. at 37°C. RNase inhibitor (160 U/ 50 μl of sample) (Invitrogen) was then added to neutralize any remaining RNase activity by incubating for 15 min. at 37°C. GII.4 RNA (1 x 10^5 PDU) was added to the reaction tube either before or after RNase treatment to check the efficacy of the RNase A or the RNase inhibitor, respectively. Ten-fold serial dilutions of stool-extracted NV RNA were used to generative a standard curve plotting the log PDU estimate vs the real-time CT value. Real-time CT values generated after the nuclease protection experiments could then be given a PDU value based on this semi-quantitative standard curve.

RESULTS

Optimization of carbohydrate magnetic separation

The carbohydrate magnetic separation (CMS) was optimized for detecting low copy numbers of Norwalk virus particles without nonspecific binding of virus to magnetic beads in the absence of carbohydrate. Norwalk virus (8FIIb 23-3b) did bind to a panel of synthetic HBGA in CMS including H type 1, H type 3, and Lewis b while no binding was detectable for H type 2 carbohydrate (data not shown). H type 1 carbohydrate was chosen for subsequent experiments because it demonstrated detection of Norwalk (NoV) virus binding at the lowest input level (20 PDU) whereas the lowest detectable input levels for H type 3 and Lewis b were 200 and 2000 PDU, respectively (n = 3) (data not shown).

Three blocking buffers were examined for their ability to block nonspecific binding to uncoated magnetic beads without interfering with specific binding of NoV to H type 1 carbohydrate or the reproducibility of results. In its original format (Harrington et al., 2004), the CMS assay used 5% Blotto as a blocking buffer. However, this blocking buffer variably interfered with specific virus binding and unreliably blocked nonspecific binding, working properly in only 4 of 7 experiments with input levels between 2,000 and 20,000 PDU. Furthermore, 5% Blotto appeared to be inhibitory to RT-PCR after heat release of the viral RNA as the 100-fold dilution of virus-bound beads was often but not always brighter (as seen on an ethidium bromide stained gel) than the 10-fold dilution or undiluted sample (4/7 experiments). When 10 % Blotto was used as a blocking buffer, inhibition of RT-PCR became more apparent as only a faint band was seen only on the 100-fold dilution of virus-bound beads and no bands were detected for the 10-fold diluted or undiluted samples (n = 1) (data not shown). The use of 5% FBS as a blocking buffer did not interfere with specific binding and blocked nonspecific binding in 4/7 experiments with a virus input level of 20 to 20,000 PDU, yet inhibition of RT-PCR was often observed for the undiluted sample but not the 10 -fold diluted sample of virus-bound beads (6/7 experiments). Although not effective at blocking nonspecific binding to uncoated beads at virus concentrations greater than 2,000 PDU, SuperBlock had the most reproducible results allowing detection of low input levels (20 to 2,000 PDU) and blocking nonspecific binding when 2,000 PDU or fewer viruses were analyzed for 8 of 8 experiments. Furthermore, SuperBlock did not inhibit RT-PCR for the undiluted bead samples for any experiment (n = 8). Therefore SuperBlock was used in the final format of the CMS.

Sensitivity and Specificity of CMS

The detection limit of CMS was in the range of 2 to 2000 NoV PDU with a median of 200 NoV PDU (range from 20 to 2000 PDU; n = 10). Neither rotavirus nor astrovirus inhibited binding of Norwalk to the H-type 1 carbohydrates. When high titers cell culture purified rotavirus group A or astrovirus were added to each reaction tube at each input level (20, 200 or 2,000 NoV PDU), the detection limit of the assay remained at 200 NoV PDU using conventional RT-PCR.

RNase treatment

To determine if bound Norwalk virus particles maintain a complete capsid, and hence are potentially infectious, we treated NoV bound to H type 1 carbohydrate with RNase A (10 μg). After neutralization of the enzyme with RNase inhibitor, bound Norwalk virus particles were heat released and viral RNA was estimated by TaqMan real-time RT-PCR. There was no significant reduction in detectable virus between the RNase treated and untreated 37°C controls for the 20,000 or 4,000 NoV PDU input levels ($p \geq 0.8$) but there was a significant difference between the RNase treated and untreated 4°C controls ($p \leq 0.02$) (Figure 2.1).

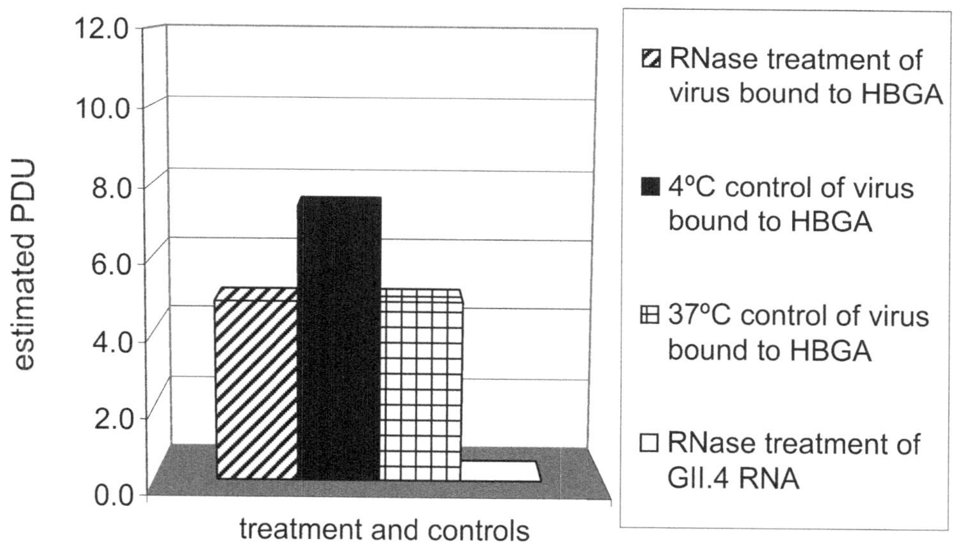

Figure 2.1 Effect of RNase on 4,000 PDU of Norwalk virus bound to HBGA and processed using CMS

However, the reduction was less than a one log from untreated to treated sample going from a mean of 41.7 (sd= 9.7) and 7.4 (sd= 2.8) to 24.8 (sd= 14.4) and 4.8 (sd= 2.8) PDU of RNA for the 20,000 and 4,000 PDU input levels. The difference in CT value between the 4°C controls and the 37°C controls was significant at the 4,000 PDU input level, but not at the 20,000 PDU input level (p = 0.03 and p = 0.07, respectively). RNase treatment of the GII.4 control RNA resulted in complete loss of a realtime RT-PCR signal.Environmental water samples concentrated by PEG precipitation were spiked with 2, 20, 200 or 2,000 NoV PDU before subjection to CMS (data not shown). The detection limit was found to be between 20 to 2000 (median = 200) for all of the water types tested (Table 2.1). NoV spiked into surface water concentrate could be detected consistently at the 20 PDU input level (n=4) and concentrate from effluent wastewater (treated sewage effluent) was consistently detected at the 200 PDU input level (n=4). The influent wastewater (raw sewage) concentrate was more inconsistent having a detection limit of 200 or 2000 PDU (n=2). While the detection limit of the CMS assay related inversely with the quality of the environmental water sample tested, the assay proved robust for detecting low levels of NoV virus contamination in environmental waters.

Table 2.1
Detection limit of CMS for NoV GI.1 in the presence of environmental water matrix as determined by number of replicates that tested positive for NoV GI.1

Input NoV PDU				Type of water
2	20	200	2000	
0/4	4/4	4/4	4/4	Surface Water
0/4	0/4	4/4	4/4	Effluent wastewater
0/2	0/2	1/2	2/2	Influent wastewater

DISCUSSION

This study provides a framework for a proof-of-concept model using a carbohydrate separation (CMS) technique for detecting infectious noroviruses in environmental waters. CMS is based on the immunomagnetic separation (IMS) technique concentrating and purifying low levels of microbes from environmental waters. IMS techniques have been developed for environmental detection of *Cryptosporidium* and *Giardia,* (USEPA, 1999), *E. coli* (Bennett et al., 1996), Hepatitis A virus (HAV) (Jothikumar et al., 1998), PV-1 (Schwab et al., 1996) and human NoVs (Gilpatrick et al, 2000, Kobyashi et al., 2004; Myrmel et al, 2000) in the context of water and foods. Downstream applications include molecular detection, immunofluorescence microscopy and cell culture. IMS uses virus-specific antibodies to concentrate viruses from larger volumes and through washing, helps to purify the viruses from inhibiting elements thereby accommodating RT-PCR based detection. Similarly, we have shown here that CMS can be used to detect low input levels of NoV RNA even in the context of other enteric viruses. However, future experiments will be needed to determine if the increasing reduction in virus detection by CMS that corresponds with decreased water quality of our environmental water samples is due to the decreasing performance of CMS or increasing sample inhibition. Furthermore, CMS can recover NoV in the context of environmental waters. Unlike the IMS, CMS incorporates the putative virus receptor for human NoV; therefore, only viruses complete in the receptor binding region will be detected. It is however not clear that this represents a real advantage. Through treatment of HBGA-bound NoV with RNase A in CMS, we determine the virus capsid is intact. Based on these results, we predict CMS can better make a distinction between infectious and noninfectious virus.

Volunteer challenge studies with Norwalk virus (GI.1) have definitively linked symptomatic Norwalk virus infection with HBGA expression (Hutson et al., 2002; Lindesmith et al., 2003). Individuals known as "secretors", express HBGAs on their gastrointestinal mucosa, its secretions, and in saliva and are genetically susceptible to symptomatic infection while "non-secretors" are resistant (Lindesmith et al., 2003). However, the role of HBGAs in one's susceptibility to symptomatic NoV infection for genogroups other than Norwalk virus has not been established as symptomatic infection with Snow Mountain virus (SMV), a GII.2 NoV, is not associated with an individual's secretor status (Lindesmith et al., 2005). The P2 region of the capsid protein is encoded by the most hypervariable region of the genome and coincides with the receptor binding region (Katayama et al, 2002; Caul et al., 2007). Changes in this region from genotype to genotype may result in changes in receptor binding. At least four HBGA binding patterns have been reported for NoVs (Huang et al., 2003) and some NoVs do not bind to any HBGA tested (Harrington et al., 2004). In vitro studies have also revealed norovirus rVLP binding to an undefined, 105-kDa, ubiquitously expressed protein designated as "norovirus attachment (NORVA) protein" and to soluble histone molecules (Histone H1) in mammalian cells (Tamura et al., 2000, Tamura et al., 2003). Membrane associated heparin sulfate proteoglycans also permit binding of norovirus rVLPs in a region distant from the HBGA binding site (Tamura et al., 2003). The role of these molecules in human NoV infectivity remains to be determined. Recently the first in vitro cell culture infectivity assay for human NoV was developed using 3-D cultures of an intestinal cell line (Straub et al., 2007). This assay may be invaluable for elucidating the role of HBGAs and other putative receptor molecules in the infectivity of NoVs.

Whether or not the HBGA binding assay can be generalized for other genogroups, particularly GII.4 NoVs most frequently associated with outbreaks of NoVs, will likely determine the applicability of CMS as a recovery method. In addition to the HBGA binding of Norwalk virus to H type 1, we have shown SMV binding to H type 3 and three GII.4 NoV strains (a Camberwell-like, Grimsby-like and a Farmington Hills-like) binding to H type 3 and/or Le^y carbohydrates adding to and confirming the results of others (Harrington et al., 2004). Thus, future experiments should be directed toward uncovering a cocktail of HBGAs that will be able to bind the majority of circulating strains of noroviruses. If successful, CMS may serve as part of a rapid detection method for noroviruses in complex matrices such as environmental waters or food. Furthermore, with validation of HBGA association with infectivity of NoV genogroups other than Norwalk by human volunteer challenge or in vitro cell culture, CMS may serve as a surrogate model for human NoV infectivity.

CHAPTER 3
RAPID AND SENSITIVE DETECTION OF NOROVIRUSES USING TAQMAN-BASED ONE-STEP RT-PCR ASSAYS AND APPLICATION TO CLINICAL AND NATURALLY CONTAMINATED SHELLFISH SAMPLES

INTRODUCTION

Noroviruses (NoV) have emerged as a leading cause of outbreaks of acute viral gastroenteritis worldwide and have been identified as the primary pathogen associated with shellfish-borne gastroenteritis (Koopmans and Duizer, 2004; Lees, 2000; Shieh et al., 2000). Since NoV is mainly transmitted through the fecal-oral route and the ID50 has been reported to be low (Moe, 2001), the risk of infection after consumption of raw or improperly cooked seafood or exposure to contaminated recreational water is considered high (Lees, 2000). Genetically NoVs can be subdivided into five separate groups or genogroups (GI, GII, GIII, GIV and GV) of which viruses of GI, GII, and GIV have been detected in humans (Koopmans et., 2000). Within the human genogroups NoV strains can be further delineated into at least 31 genetic clusters or genotypes based on the complete capsid protein VP1 (Zheng et al., 2006). Although reverse transcription-polymerase chain reaction (RT-PCR) has become the standard for diagnosis of NoV infection worldwide (Atmar and Estes, 2001, Fankhauser et al., 2002; Vinjé et al., 2003), many laboratories use multiple sets of primers targeting different regions of the genome because of the difficulty to design broadly reactive primers that are able to sensitively detect all the different NoV genotypes (Vinjé et al., 2003). In addition, confirmatory steps are needed to evaluate the specificity of the amplification product (Fankhauser et al., 2002; Vinjé et al., 1996; Vinjé and Koopmans 2000), or to increase the sensitivity for the detection of NoV in food and water matrices (Green J, et al., 1998, Lamothe et al., 2003). Recently, both SYBR Green as well as TaqMan-based RT-PCR assays for the detection of NoV in clinical specimens have been reported (Bhattacharya et al., 2004; Hohne and Schreier 2004; Kageyama et al., 2003; Pang et al., 2004). However, none of these assays used combined RT and PCR reactions in one tube, were able to detect both GI and GII in one reaction, or were validated with naturally contaminated environmental (shellfish) samples. One of the TaqMan-based assays targeted a region just downstream of the region B primers (Fankhauser et al., 2002) near the ORF1-ORF2 junction of the genome, which had been shown to be the most conserved region of the NoV genome (Katayama et al., 2002). Crucial for the sensitive detection of NoV in environmental samples like shellfish is a rapid assay with a broad reactivity as well as a low detection limit. Real-time PCR may be considered to be preferable due to its simplicity and ability to generate quantitative data. In this chapter, we report the development of a one-step TaqMan® RT-PCR assay for the sensitive detection of human NoV strains and its validation using panels of archived stool specimens and naturally contaminated shellfish samples.

MATERIALS AND METHODS

Clinical strains and shellfish samples

A panel of 84 stool specimens collected over a 5-year period of time (1996–2000) was derived from both outbreaks and sporadic cases of gastroenteritis. These specimens were tested previously for NoV by RT-PCR using primers targeting conserved regions of the ORF1 that include the GLPSG motif of the RNA-dependent RNA polymerase (Jiang et al., 1999; Vinjé and Koopmans 1996) or region B (Fankhauser et al., 2002). The specimens had been stored at -80°C and represented most of the NoV genotypes that had been recently circulating in Europe and the USA (Table 3.1). In addition, eight specimens containing other enteric viruses (adenovirus type 40 and type 41, echovirus 11, rotavirus strain SA-11, hepatitis A virus strain HM175, poliovirus Sabin type 1) were included in this study. Stool samples were diluted to 10% in PBS, vortexed and clarified by centrifugation at 2,000 x g for 15 min.

All shellfish used in this study were collected from commercial harvesting areas in the north Atlantic region during 2002-2004. The numbers of samples of each species used were as follows: 36 Pacific oysters (*Crassostrea gigas)*, and 2 native oysters (Ostrea *edulis).* For each sample a minimum of six shellfish from the same species were aseptically opened and the animals were removed from their shells. The peripheral flesh and organs of each animal were then cut away from the hepatopancreas which were then finely chopped as described previously (Henshilwood et al., 2003). To degrade shellfish tissue and allow release of virions into solution an equal volume of 100 µg/ml Proteinase K solution was added and the the sample was then incubated at 37°C for 1 hr. with shaking at 320 rpm. The reaction mixture was subsequently incubated at 65°C for 15 min. in order to inactivate the proteinase and the soluble portion (shellfish concentrate) was collected by centrifugation at 3000 x g for 5 min, and stored at 4°C.

Viral RNA extraction

For stool samples, viral RNA was extracted from clarified 10% extracts in PBS using QIAamp microspin columns (viral RNA mini kit; Qiagen, Valencia, CA) according to the manufacturers' instructions and stored at –80°C. For each shellfish sample, one aliquot each of 400 µl and 133 µl shellfish homogenate for nested PCR and 1-3 aliquots of 133 µl shellfish homogenate for real-time PCR analysis were added to separate 1.5 ml microcentrifuge tubes containing 10 µl silica bead suspension ('glassmilk': Anachem Ltd, Luton, UK). Final volumes were adjusted to 410 µl by adding PBS (Oxoid, Basingstoke, UK). Viral RNA from purified shellfish concentrates was extracted by binding to size-fractionated silica after treatment with guanidine isothiocyanate (GITC) as described previously (Boom et al., 1990; Lees et al., 1994). After eluting from the silica particles, RNA was precipitated in ethanol and sodium acetate. The RNA pellet was resuspended in 8.9 µl of reverse transcription (RT) reaction mix and immediate subjected to reverse transcription as described previously (Green J et al., 1998).

Table 3.1
Panel of norovirus positive stool samples used in this study

Genotype	Prototype [a]	strain ID
GI.1	Norwalk	8FIIb
GI.2	Southampton	c2, S7, sw8
GI.3	Desert Shield	c3, S10, sw6, sw18
GI.4	Chiba	B39, c4, S11, sw1, sw2
GI.5	Musgrove	c5
GI.6	Hesse	B18, B49, sw3
GI.7	Winchester	c7
GII.1	Hawaii	sw7, R4, R18, sw16, c8, G4, CDC8
GII.2	Melksham	SMV1-4, S2, R10, B13, B16, c9, sw5, CDC9
GII.3	Toronto	B2, B3, B40, G4, c10, S13, R8, R9, R21, J8b, J21a, J21b, J24, sw10, sw13, sw14
GII.4	Bristol	B4, B26, B27, B28, B45, B48, B51, S19, sw17, sw19, sw20, sw22, sw23, sw24, G1, G2, c11
GII.5	Hillingdon	c12, CDC12
GII.6	Seacroft	c13, G3, R16, B11, J11, R1, sw11, sw12, CDC13
GII.7	Leeds	c14, J9a, M13, CDC14
GII.8	Amsterdam	B5
GII.9	Virginia	J10b
GII.12	Wortley	M17, J1a, J15a
GII.14	Fayetteville	M12, M27
GIV.1	Alphatron	c17

[a] Zheng et al., 2006

Source: Jothikumar, N., Lowther, J.A., Henshilwood, K., Lees, D.N., Hill, V.R., and J. Vinjé. 2005. Rapid and sensitive detection of noroviruses by using TaqMan-based one-step reverse transcription-PCR assays and application to naturally contaminated shellfish samples. *Appl. Environ. Microbiol.* 71(4):1870-5. With permission from American Society for Microbiology.

Reverse transcription and nested PCR for shellfish

cDNA was generated using random hexamers and first round PCR was performed using G1-G2 and SM31 primers (Green J et al., 1998) followed by second-round amplification using 1-μl of the first-round amplicon and 49 μl of PCR mix containing 10 mM Tris-HCl pH 8.8, 50 mM KCl, 1.25 mM $MgCl_2$, 0.2 mM of each dNTP, 1 unit AmpliTaq Gold enzyme (ABI), and 20 pmol of each primer (Ando and E3 for GI and NI and E3 for GII). The reaction mixes were then analyzed using agarose gel electrophoresis. RT-PCR products of the expected size (GI and GII primers amplify 116 bp and 114 bp products respectively) were cloned and sequenced by using previously described methods (Henshilwood et al., 1998). Phylogenetic analysis was performed using Lasergene software (DNAStar). NoV sequence in either of the two reactions per shellfish sample was counted as an overall positive for that sample.

Table 3.2
Oligonucleotides for Taqman®-based NoV RT-PCR used in this study

Genogroup	Oligo ID	DNA sequence (5'-3') [b]	Orientation	Location [c]
GI	JJV1F	GCC ATG TTC CGI TGG ATG	+	5282 -
	JJV1R	TCC TTA GAC GCC ATC ATC AT	–	5377 -
	JJVIP	FAM-TGT GGA CAG GAG ATC GCA ATC TC-BHQ	+	5319 -
GII	JJVIIF	CAA GAG TCG ATG TTT AGG TGG ATG AG	+	5012 -
	COGIIR [a]	TCG ACG CCA TCT TCA TTC ACA	–	5100 -
	RINGII-TP [a]	FAM-TGG GAG GGC GAT CGC AAT CT-BHQ	+	5048 -

[a] Kageyama et al., 2003
[b] I = deoxyinosine, in duplex assays a 5'-FAM labeled GI probe was combined with a 5'-JOE or 5'-TET labeled GII probe. BHQ = Black Hole Quencher
[C] Nucleotide positions based on Norwalk [GI]- M87661, Lordsdale [GII]- X86557 and FtLauderdale [GIV]- AF414426.

Source: Jothikumar, N., Lowther, J.A., Henshilwood, K., Lees, D.N., Hill, V.R., and J. Vinjé. 2005. Rapid and sensitive detection of noroviruses by using TaqMan-based one-step reverse transcription-PCR assays and application to naturally contaminated shellfish samples. *Appl. Environ. Microbiol.* 71(4):1870-5. With permission from American Society for Microbiology.

TaqMan® RT-PCR primer and probe design

To identify conserved regions across the NoV genome that are suitable for the development of real-time RT-PCR primers and probes, NoV sequences from databases available at the University of North Carolina at Chapel Hill (UNC), CEFAS, and GenBank were obtained. Several regions in ORF1 as well as ORF2 were considered and based on multiple alignments several candidate primers and probes were selected for evaluation including recently published TaqMan® reagents (Kageyama et al., 2003). Although initially several promising primer sets targeting region C (Kojima et al., 2002; Noel et al., 1997) and region D (Vinjé et al., 2004) were identified and evaluated, we selected genogroup-specific primers and probes targeting the conserved region at the ORF1-ORF2 junction of the genome (Table 3.2). We used Metcalf software (Schutz and von Ahsen 1999) to design consensus primer/probe sequences by calculating the Tm of the primer with each individual NoV sequence used in this study. Care also was taken to avoid any primer-primer self-annealing. Primer JJV1F included a single deoxyinosine at position 12 to compensate for a four-fold degeneracy. In its final format, the GI primers (JJV1F and JJV1R) amplified a 96-bp fragment. A TaqMan® probe (JJV1P) was designed to detect all GI strains. The GII primers amplified an 89-bp fragment and the forward primer (JJV2F) was designed to hybridize to all genetic clusters of genogroup II. FAM (6-carboxyfluorescein)-5'-labeled probes were used for monoplex assays whereas in duplex assays a FAM-labeled GI and a JOE (6-carboxy-4',5'-dichloro-2',7'-dimethoxyfluorescein) or HEX (6-carboxy-2',4,4',5',7,7'-hexachloro-fluorescein) labeled GII probe were used. The probes contained either the 3' Black Hole quencher (BHQ) dye (biotechnology core facility at CDC-Atlanta) or TAMRA (Applied Biosystems) for the analyses of shellfish.

TaqMan® RT-PCR assays

For two-step TaqMan® assays, random hexamers (ABI) were compared to specific reverse primers using the GeneAmp RNA PCR Core kit (Applied Biosystems). RT was performed at 42°C for 15 min., followed by incubation at 99°C for 5 min. and cooling to 5°C for 5 min. in GeneAmp PCR system 9700 (Applied Biosystems). TaqMan® PCR was then performed using the QuantiTect Probe PCR kit (Qiagen).

In its final format, the reaction volumes contained 0.5 or 1 µl of template RNA (stool samples) or 2 µl cDNA (shellfish samples), 250 nM of each primer, 100 nM of JJV1P probe (for GI viruses), or 100 nM of RING2-TP probe (for GII viruses) (Kageyama et al., 2003), and QuantiTect™ Probe RT-PCR buffer (Qiagen, Valencia, CA) for stool samples or TaqMan® Universal PCR Master Mix (Applied Biosystems) for shellfish samples. For stool samples, the reaction mixture was then subjected to a one-step format according the following amplification conditions: (1) RT step for 30 min. at 55°C, (2) 15 min. at 95°C to activate Taq polymerase, and (3) 45 cycles of 10 s at 94°C, 20 s at 55°C, and 15 s at 72°C. cDNA from shellfish samples was amplified as follows: 2 min. at 50°C, then 10 min. at 95°C followed by 50 cycles of 15 s at 95°C and 1 min. at 60°C. Several thermocyclers were used to perform the TaqMan® RT-PCR assay developed in this report: iCycler iQ™ (Bio-Rad, Hercules, CA), R.A.P.I.D. Cycler (Idaho Technology Inc, Salt Lake City, UT), and SmartCycler (Cepheid, Sunnyvale, CA) for clinical samples, and ABI GeneAmp SDS 5700 real-time PCR machine (Applied Biosystems, Foster City, CA) for shellfish samples.

All amplification reactions were carried out in duplicate. To determine the correct product size and to compare the sensitivity with fluorescent-tagged detection, RT-PCR products of TaqMan® reactions from clinical samples were visualized on ethidium bromide-stained agarose gels. Shellfish samples giving a positive result in either or both of the duplicate reactions were counted as an overall positive.

Standards for Real-Time RT-PCR assay

To obtain representative positive control standards, amplicons generated from strains S11 (GI.4) and SMV1-4 (GII.2) using the GI and GII primers, respectively, were cloned into TrueBlue® MicroCartridge™ PCR Cloning Kit (Genomics One, Quebec) and plasmid DNA was purified using the Nucleobond 100 kit (Promega). The concentration of each purified plasmid was quantitated by spectroscopy at 260 nm and 10-fold serial dilutions of each plasmid DNA were used for the preparation of the standard curve.

Discrepancy testing

Discrepancy testing was performed when any shellfish cDNA sample resulted in a discrepant TaqMan® result compared to the corresponding nested PCR result for the same genogroup. Duplicate 2-µl aliquots of cDNA used for the TaqMan® analysis were analyzed by nested PCR (Green J et al., 1998). Samples giving a single positive duplicate reaction were counted as an overall positive and the result superseded the original nested PCR result if different, for the purposes of comparative analysis.

RESULTS

Specificity of genogroup-specific TaqMan® RT-PCR assays

For the detection of NoV in clinical samples, we chose a one-step RT-PCR approach. For shellfish extracts, random primer generated cDNA was used, and hence a two-step approach was followed. The individual GI and GII TaqMan® RT-PCR assays were validated using a panel of 84 stool specimens containing NoV-RNA as detected by conventional RT-PCR (Vinjé and Koopmans 1996; Vinjé et al., 2004). All RT-PCR products had the appropriate size when analyzed on ethidium bromide stained agarose gels (data not shown). The GI and GII TaqMan® assays were genogroup specific and no cross-reactions with strains from other genogroups were observed. Based on these results we designed a one-step GI/GII duplex assay for sensitive detection as well as distinguishing the most prevalent genogroups. In its optimized duplex format we use 5'-FAM labeled GI probe and a 5'-JOE labeled GII probe.

Limit of detection of NoV TaqMan® assays

The sensitivity of the monoplex assays was determined using 10-fold serial dilutions of NoV GI and GII plasmids (Figure 3.1). To determine the sensitivity, of the real-time RT-PCR assays, standard curves of crossing point (Ct) values versus number of copies were generated by analyzing four replicates of 10-fold serial dilutions (7.2×10^6 to 7.2×10^0 copies) of NoV-GI plasmid and (5.9×10^6 to 5.9×10^0 GII.2 copies) of NoV-GII plasmid. In all cases, excellent correlation between amount of target template and Ct value was obtained and the sensitivity of both NoV real-time RT-PCR assays was calculated as <10 copies of viral genome copies/reaction.

Parameters of NoV TaqMan® RT-PCR assays

Several different variables of the assay (random priming versus specific primer for RT, 2-step versus one-step format, and monoplex versus duplex (GI and GII combined)) were compared. Using a panel of 8 different strains (4 GI, 4 GII), Ct values were lower (range 2.2 - 7.8) when specific primers were used compared to random primers whereas the differences between 2-step format and 1-step format were less than 1 Ct value (data not shown). A total of 25 NoV clinical samples representing strains from all genotypes available in this study was analyzed by duplex and monoplex assays. The duplex assay (GI_FAM labeled; GII_JOE labeled) showed comparable sensitivity compared to monoplex assay (FAM labeled probe) with average Ct values of the duplex assay less than 1.0 higher than the monoplex assay (data not shown).

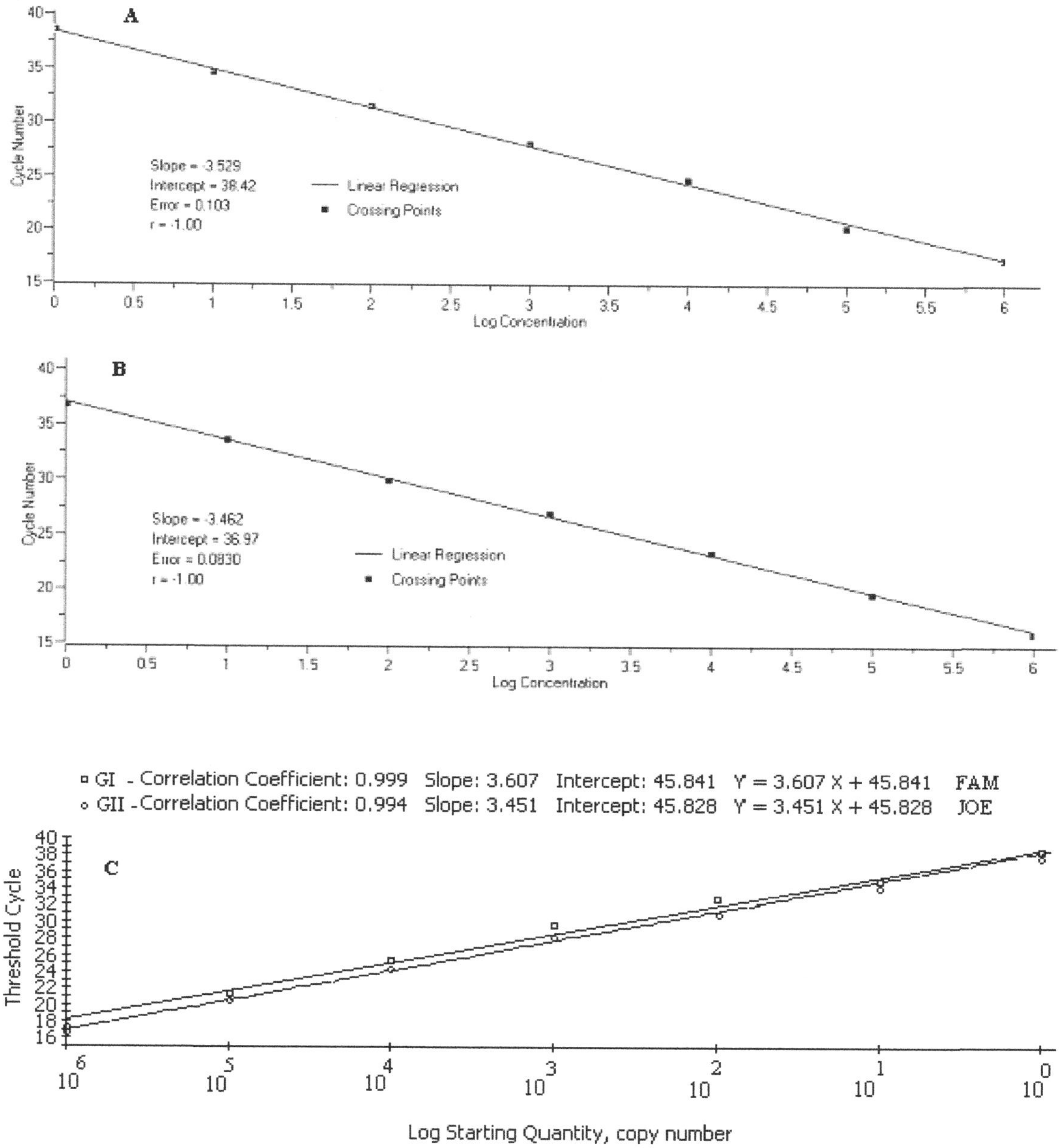

Figure 3.1 Standard curves for GI.4 and GII.4 plasmid standards containing from 1 to 10^6 copies per reaction. (A) Detection limit for the NoV GI TaqMan assay based on a dilution series of the GI.4 plasmid with the R.A.P.I.D. Cycler. (B) Detection limit for the NoV GII TaqMan assay based on a dilution series of the GII.4 plasmid with the R.A.P.I.D. Cycler. (C) Detection limit for the NoV duplex TaqMan assay based on a dilution series of the GI.4 and GII.4 plasmids with the Bio-Rad iCycler iQ. Source: Jothikumar, N., Lowther, J.A., Henshilwood, K., Lees, D.N., Hill, V.R., and J. Vinjé. 2005. Rapid and sensitive detection of noroviruses by using TaqMan-based one-step reverse transcription-PCR assays and application to naturally contaminated shellfish samples. *Appl. Environ. Microbiol.* 71(4):1870-5. With permission from American Society for Microbiology.

Table 3.3
Comparision of Ct values of UNC-NoV and Kageyama NoV TaqMan® assays from 20 panel strains

Strain	Genotype	NoV TaqMan RT-PCR assay: This study	Kageyama et al., 2003	Ct difference
Sw8	(GI.2)	23.7	28.2	4.5
Sw6	(GI.3)	23.8	26.0	2.2
Sw18	(GI.3)	24.2	32.3	8.1
Sw1	(GI.4)	23.8	30.4	6.6
Sw3	(GI.6)	28.8	36.1	7.3
B18	(GI.6)	29.2	38.1	8.9
Sw17	(GII.1)	18.6	23.8	5.1
CDC8	(GII.1)	24.8	27.6	2.8
CDC9	(GII.2)	23.5	27.6	4.1
R10	(GII.2)	29.0	34.1	4.9
S13	(GII.3)	26.6	35.1	8.5
B40	(GII.3)	22.9	29.1	6.2
R9	(GII.3)	25.0	29.1	4.1
B27	(GII.4)	23.9	25.9	2.0
Sw10	(GII.4)	29.9	33.9	4.0
B4	(GII.4)	27.6	30.0	2.4
CDC12	(GII.5)	29.3	33.9	4.6
CDC13	(GII.6)	35.1	37.2	2.1
R16	(GII.6)	21.2	22.8	1.6
CDC14	(GII.7)	28.3	33.8	5.5

Source: Jothikumar, N., Lowther, J.A., Henshilwood, K., Lees, D.N., Hill, V.R., and J. Vinjé. 2005. Rapid and sensitive detection of noroviruses by using TaqMan-based one-step reverse transcription-PCR assays and application to naturally contaminated shellfish samples. *Appl. Environ. Microbiol.* 71(4):1870-5. With permission from American Society for Microbiology.

Comparison of NoV TaqMan® assays with recently published NoV TaqMan assays

To compare the sensitivity of our monoplex NoV TaqMan® assays with the recently published assay (Kageyama et al. 2003), we analyzed a panel of 17 representative strains using a one-step RT-PCR format and determined the Ct values for the samples. Overall, our monoplex real-time assay was 10 to 100-fold (4.12 ± 2.16 lower Ct values) more sensitive than the other reported assays defined by limit of detection (Table 3.3). In parallel, a small panel of samples was also analyzed by another NoV TaqMan® assay (Hohne et al., 2004), but a full comparison was not performed because this assay did not detect any of the GI strains (data not shown). The only difference between our GII assay and the assay described by Kageyama et al. is the forward primer (which Kageyama et al. designed with several degeneracies, but which we designed with no degeneracies).

To evaluate the robustness of our assay versus Kageyama et al. 2003, we decided to compare both assays by gradient real time RT-PCR. The Ct versus temperature profiles for a panel of five different GII strains clearly show that primer JJV2F yields consistent low Ct values across the entire range (54–62°C) of annealing temperatures tested, whereas the COG2F primer yielded increasingly higher Ct values as annealing temperatures increased, and no fluorescence signal at 60°C and higher (Figure 3.2).

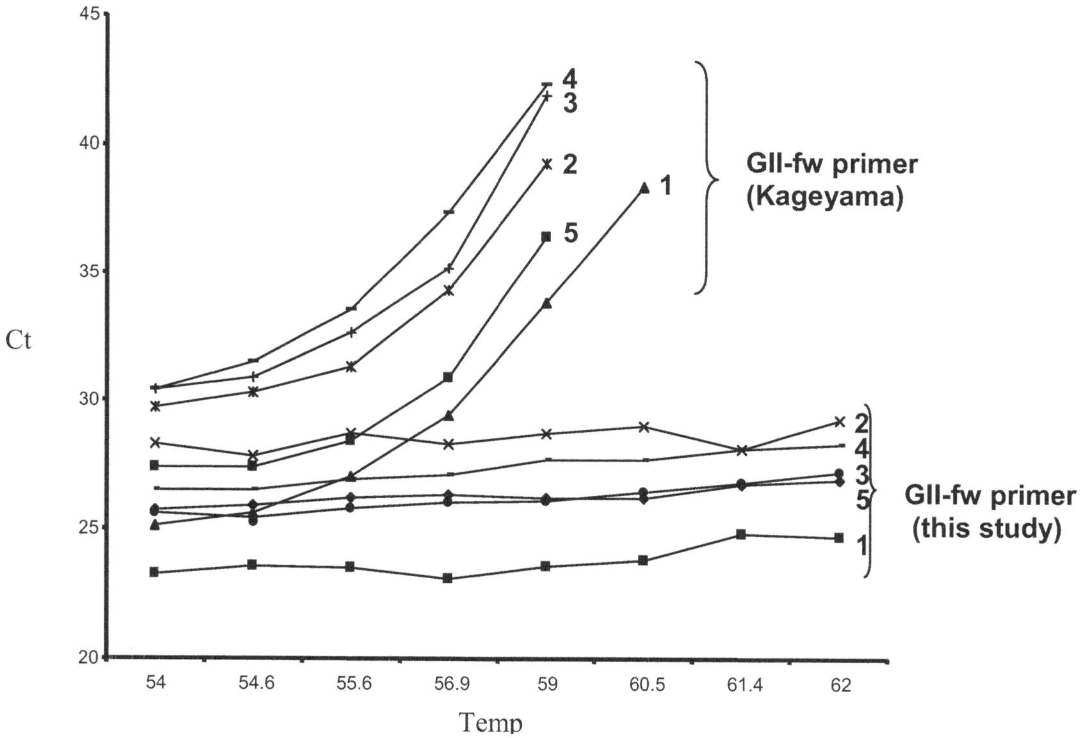

Figure 3.2 Comparison of Ct values for two different GII forward primers, JJV2F (this study) and COG2F (Kageyama et al., 2003), determined by gradient TaqMan RT-PCR with RNA from five different GII strains (lines 1, R4 [GII.1]; lines 2, c8 [GII.1]; lines 3, c9 [GII.2]; lines 4, B2 [GII3]; lines 5, sw12 [GII.6]). Source: Jothikumar, N., Lowther, J.A., Henshilwood, K., Lees, D.N., Hill, V.R., and J. Vinjé. 2005. Rapid and sensitive detection of noroviruses by using TaqMan-based one-step reverse transcription-PCR assays and application to naturally contaminated shellfish samples. ***Appl. Environ. Microbiol.*** **71(4):1870-5. With permission from American Society for Microbiology.**

Detection of NoV in shellfish extracts

Eighteen of the 38 shellfish extracts tested positive for GI-NoV by both nested PCR and TaqMan® RT-PCR (Table 3.4). In total, 6 extracts showed discordant results with 4 samples testing only positive by TaqMan® RT-PCR and 2 samples only positive by nested PCR. Of these 38 extracts 19 were positive for GII by both nested and TaqMan® RT-PCR. Five TaqMan® RT-PCR positive samples could not be confirmed by nested PCR and one sample tested positive by nested PCR and negative by TaqMan® RT-PCR (Table 3.4). Nested PCR products were sequenced and genotyped (data not shown).

Table 3.4
Comparison of performance of TaqMan RT-PCR assays and nested PCR for the detection of norovirus RNA in 38 shellfish extracts (A: GI, B: GII)

A

		TaqMan RT-PCR	
		pos	neg
Nested PCR	pos	18	2
	neg	4	14

B

		TaqMan RT-PCR	
		pos	neg
Nested PCR	pos	19	1
	neg	5	13

Source: Jothikumar, N., Lowther, J.A., Henshilwood, K., Lees, D.N., Hill, V.R., and J. Vinjé. 2005. Rapid and sensitive detection of noroviruses by using TaqMan-based one-step reverse transcription-PCR assays and application to naturally contaminated shellfish samples. *Appl. Environ. Microbiol.* 71(4):1870-5. With permission from American Society for Microbiology.

Ct values of shellfish samples were at, or near, the limit of assay sensitivity and frequently were positive in only 1 of 2 replicates. For the GI assay, 55% of positive samples were positive in only one replicate with Ct values ranging from 39-43. However, for the samples testing positive in both duplicates the majority (80%) gave Ct values of <40. For the GII assay, Ct values ranged from 38-46. In comparison to the GI real-time assay fewer samples were positive in only one duplicate (25% of positive samples) but a lower proportion of samples gave average Ct values <40.

DISCUSSION

Noroviruses (NoVs) are the most common cause of foodborne outbreaks of acute gastroenteritis and contaminated shellfish is one of the most important food items involved in such outbreaks (Koopmans et al., 2002; Parashar and Monroe 2001; Shieh et al., 2000). Hence, introduction of routine norovirus monitoring of shellfish would be highly advisable but most published conventional RT-PCR assays using a single primer pair are unable to detect all human noroviruses (Vinjé et al., 2003). Although nested PCR assays have been successfully employed for the amplification of low levels of virus found in naturally contaminated shellfish (Green J et al., 1998; Lees et al., 1995; LeGuyader et al., 2003; Myrmel et al., 2004; Nishida et al., 2003; Shieh et al., 2003), no International Organization for Standardization (ISO) method currently exists and to our knowledge no laboratory offers an accredited service for detection of these viruses.

In the present study we have designed and evaluated broadly-reactive one-step TaqMan-based RT-PCR assays for the sensitive detection of genogroup I and II NoVs. The assays were validated against the full range of NoV genotypes from clinical samples and shown to be at least as sensitive and broadly reactive as conventional nested PCR for detection of norovirus in shellfish samples. Initially several promising generic primer sets targeting region C (Kojima et al., 2002; Noel et al., 1997) and region D (Vinjé et al., 2004) of GI and GII NoV were identified and evaluated for real-time RT-PCR (data not shown), but in the final protocol, the ORF1-ORF2 junction region was selected as the target region because this region has been shown to have the highest similarity across the genome (Kageyama et al., 2003).

To balance the maximum number of mismatches, the GII forward primer JJV2F was designed in such a way that it contained intentionally introduced ambiguities towards the target sequence of different GII strains. This was accomplished by calculating the Tm for the interaction of both primers with each NoV genotype. The primer shares at least 70% homology to the variety of different GII genotype target sequences while the eight bases at the 3' end of the primer were 100% complementary to the GII genomic sequences. Because of the mismatches in the forward primer, we chose a relatively low annealing temperature (55 °C) rather than 60 °C to guarantee binding to the target sequences from stool specimens. At this lower annealing temperature, we did not observe any nonspecific amplification when analyzing the TaqMan® RT-PCR products on agarose gels (data not shown). The success of this approach was demonstrated when comparing our assays with a recently published NoV TaqMan® assay (Hohne and Schreier 2004; Kageyama et al., 2003) showing lower Ct values across the range of different genotypes. The gradient real-time RT-PCR data demonstrate the increased robustness of our GII assay using the forward primer without degeneracy, further supporting other data in this study that indicate that the sensitivity of our GII assay is as good or greater than other published GII NoV assays. In addition, our data show that a one-step duplex format is able to detect GI NoV or GII NoV in clinical samples without substantial loss of sensitivity compared to the monoplex format. However, because the level of NoV contamination in shellfish is close to the limit of detection, the performance of the duplex assay needs to be further evaluated on these samples.

There are only few published studies that have used TaqMan technology for the detection of NoVs in naturally contaminated shellfish samples (Loisy et al., 2005; Nishida et al., 2003) both of which use degenerate primers and probes that were reported previously (Kageyama et al., 2003). The increased sensitivity of our assay in terms of Ct values compared with other real-time RT-PCR assays is likely due to the selection of GII primers/probes without degeneracies and novel GI reagents. Compared to conventional multiplex RT-PCR assays, the monoplex and duplex TaqMan® RT-PCR assays reported here are more rapid because they do not require additional nested amplification and/or additional confirmation steps like hybridization or sequencing of the RT-PCR products. Identifying a possible source of a NoV outbreak related to shellfish consumption is crucial for early intervention and control and we are currently investigating the possible options to developing multiplex assays for rapid and reliable genotyping of NoV targeting region D of the capsid gene (Vinjé et al., 2004). Although an RT-PCR positive signal should be interpreted with caution since molecular methods cannot distinguish between infectious and inactivated virus, recent inactivation studies on hepatitis A virus showed a clear correlation between the detection of viral genome with the presence of infectious virus in the sample (Beuret 2004). In addition, by incorporating assays for the detection of other enteric viruses like rotavirus and astrovirus, the multiplex detection of four different viruses in one reaction would be a logical next step. The development of such TaqMan® multiplex assays would have a tremendous impact on clinical diagnosis of enteric virus infections, saving substantial time and costs when such an assay is coupled with automated nucleic acid extraction.

CHAPTER 4
COMPARISON OF CONVENTIONAL AND TAQMAN® REALTIME RT-PCR ASSAYS FOR THE DETECTION OF HUMAN NOROVIRUSES IN ENVIRONMENTAL WATER SAMPLES

INTRODUCTION

Food and waterborne diseases are a major cause of morbidity and hospitalization worldwide. Each year it is estimated that 76 million foodborne illnesses occur resulting in more than 300,000 hospitalizations and 5,000 deaths in the US alone. This indicates that 1 in 4 Americans develop foodborne illness each year and at least 1 in 1000 is hospitalized (Butt et al., 2004). Viral agents such as hepatitis A virus, rotavirus, astrovirus and especially noroviruses (NoVs) are estimated to account for approximately 50% of foodborne illnesses (Mead et al., 1999; Widdowson et al., 2005).

Viruses enter the human body predominantly via the oral route although the aerosols that are generated during explosive vomiting have also been implicated to cause outbreaks. The clinical symptoms of a norovirus infection include diarrhea, vomiting, nausea, abdominal cramps, fatigue and general malaise, although not all symptoms may be present in all infected individuals. The average incubation time is between 12 to 24 hrs and in most cases the disease is self-limiting and lasts from 12 to 60 hrs but may last longer in young children and immune compromised individuals such as the elderly. The infectious dose (ID_{50}) of norovirus is estimated to be between 10-100 infectious particles based on human volunteer studies with the prototype Norwalk virus (Moe et al., 1999). Hence, sensitive methods that are able to detect low numbers of viruses are critical in helping to control and prevent outbreaks.

In recent years, molecular detection methods (i.e., RT-PCR based methods), commonly used in clinical microbiological laboratories, have been widely implemented for the detection of pathogens in the environment. One of the biggest challenges for the detection of low-copy numbers of pathogens in environmental matrices is the presence of inhibitors of (RT)-PCR that may be co-concentrated during sample preparation. Important advantage of PCR-based methods is that they do not only provide sensitive detection of the virus but, by sequencing of the (RT)-PCR products, also allow the ability to compare strains of a possible environmental source (e.g., contaminated food or water) with clinical strains associated with the outbreak.

The genetic diversity among NoV strains makes the design of broadly reactive primers a challenge as more than 29 genetic clusters of norovirus have been identified (Zheng et al, 2006). These clusters have been assigned to five genogroups of which strains of GI, GII and GIV are found in humans. Many published primer sets are often located within a few nucleotides of each other in regions that are relatively conserved. These regions include region A (a small region of the RNA polymerase gene), region B (a region at the 3'-end of ORF1), region C (a region at the 5'-end of ORF2), and region D (at the 3'-end of ORF2). Few studies have focused on the evaluation of the sensitivity and detection limits of existing primer sets. The present study evaluates multiple published primer sets targeting the ORF1-ORF2 junction region as well as the 5'-end of ORF2 (region C) for the sensitive and broad detection of human NoV and its application to detect NoVs in naturally contaminated waste and drinking water samples.

MATERIALS AND METHODS

Selection of oligonucleotide primers

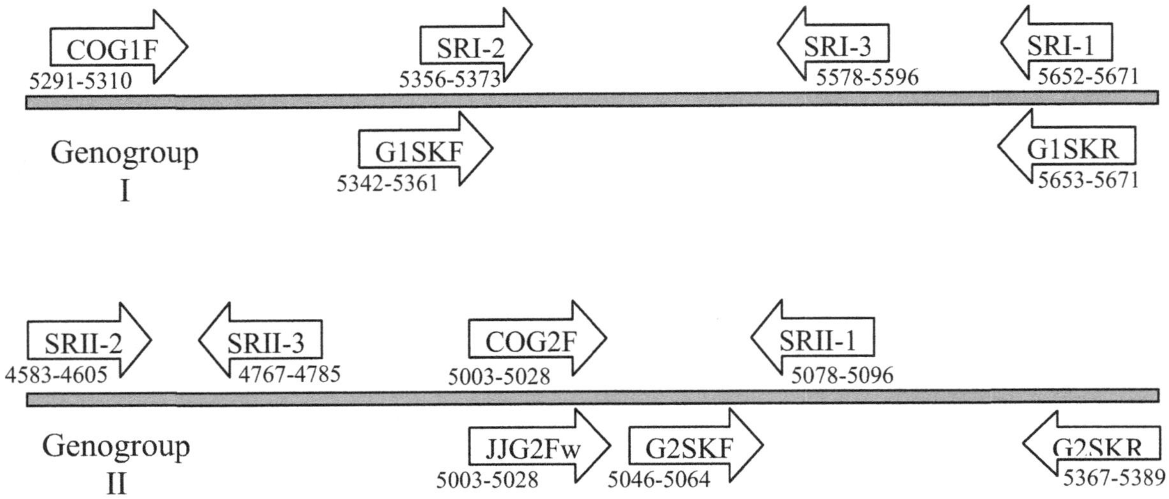

Figure 4.1 Relative primer positions on the GI (top) and GII (bottom) norovirus ORF1-ORF2 junction region used in this study.

The ORF1-ORF2 junction region has recently been established as the most conserved region across the norovirus genome (Kageyama et al., 2003). In addition, a region close to the 5'-end of ORF2 (region C) has been shown to be suitable for strain differentiation (Kojima et al., 2002; Kageyama et al., 2004). A number of published hybrid primer pairs targeting these regions were included in our initial evaluation (Hafliger et al., 1997; Kojima et al., 2002; Kageyama et al., 2003; Figure 4.1). Thirteen primer combinations, seven directed to GI and six directed to GII NoVs were selected for study (Table 4.1). In addition, we included a primer set targeting region A combining the recently published primers set (RegA – MJV12; Vinjé et al., 2004) with a hemi-nested primer (Mp290; Myrmel et al., 2004).

Table 4.1
Published and hybrid oligonucleotide primer combinations selected for evaluation

	Set	Forward	Reverse	Product Size (bp)
GI	A	G1SKF	G1SKR	329
	B	COG1F	G1SKR	380
	C	G1SKF	SRI-1	329
	D	G1SKF	SRI-3	235
	E	SRI-2	SRI-1	315
	F	SRI-2	G1SKR	315
	G	SRI-2	SRI-3	240
GII	H	G2SKF	G2SKR	343
	I	COG2F	G2SKR	386
	J	COG2F	SRII-1	93
	K	SRII-2	SRII-3	202
	L	SRII-2	SRII-1	315
	M	SRII-2	G2SKR	806

Virus strains and RNA extraction

A panel of NoV strains (n=24) representing both GI genotypes (GI.1, GI.2, GI.3, GI.4 and GI.6) and GII genotypes (GII.1 through GII.7) were selected from archived samples of various outbreaks of gastroenteritis and each sample had previously tested positive by RT-PCR targeting region A of the NoV genome (Vinjé et al., 2004; Table 4.2). Viral RNA was extracted from 10% fecal suspensions in PBS by lysis of the viral particles using guanidinium isothiocyanate and purification of the RNA by QIAamp spin columns (Viral RNA Mini Kit; Qiagen, Valencia, CA). Ten-fold serial dilutions of NoV RNA were then prepared in TE buffer and either tested directly or stored at -80°C.

Table 4.2
NoV strains used to evaluate published and hybrid primer combinations (panel 1) and hemi nested primer combinations (panel 2)

Panel #1

GI NoV	strain	genotype	strain	genotype
	Sw8	GI-2	OB35	GII-4
	CDC2	GI-2	G1	GII-4
	S7	GI-2	Sw24	GII-4
	B39	GI-4	CDC11	GII-4
	Sw1	GI-4	CDC14	GII-7
	CDC4	GI-4	M13	GII-7
	S11	GI-4	OB53	GII-7
			J9a	GII-7

Panel #2

GII NoV	strain	genotype	strain	genotype
	CDC 1	GI-1	Sw 7	GII-1
	Sw 8	GI-2	B13	GII-2
	CDC 2	GI-2	B40	GII-3
	S7	GI-2	OB35	GII-4
	Sw 6	GI-3	G1	GII-4
	Sw 18	GI-3	Sw 24	GII-4
	B39	GI-4	CDC11	GII-4
	Sw 1	GI-4	Sw 9	GII-5
	CDC 4	GI-4	Sw 11	GII-6

Environmental water samples

One liter of both influent and effluent waste water samples and five liters of drinking water influent were collected monthly (September 2003 through September 2004) from two drinking water and two waste water treatment facilities in Georgia. In addition, one-hundred liters of treated drinking water was filtered on-site through a 1-MDS filter (Cuno Inc., Meriden, Connecticut). All water samples were shipped on ice to our laboratory within 24h of collection.

Virus concentration by polyethylene glycol (PEG) concentration

Immediately upon arrival the non-filtered samples (drinking water effluent and both influent and effluent waste water) were centrifuged for 15 min. at 7280 x g. The supernatant was removed and adjusted to pH to 7.0. The pellet was resuspended in 5-10 ml of PBS and chloroform extracted. After centrifugation (2000 x g for 15 min.), the supernatant was combined with the pH adjusted supernatant. PEG-8000 (Sigma, St. Louis, MO) and NaCl were added to a final concentration of 8 % (w/v) PEG and 0.3 M NaCl and incubated overnight at 4°C to precipitate the viruses. Viruses were collected by centrifugation at 7280 x g for 60 min. After chloroform extraction of the resuspended PEG pellet, a second PEG precipitation was performed followed by chloroform extraction. After separation of the phases by centrifugation the supernatant was stored at -80°C.

For the treated drinking water, the viruses were eluted from the 1-MDS filters using 15 ml of 1.5 % (w/v) beef extract (Difco, No. 0115-17-3) in 0.05 M glycine, pH 9.5 for 10 min. at room temperature. Then the filter was incubated for 10 min. with 15 ml of 1.5% (w/v) beef extract in glycine 0.05 M, pH 7.0. The two beef extract solutions were combined (30 ml) and the pH was adjusted to pH 7.0. The viruses were then PEG precipitated as described above.

Sample RNA extraction of environmental water samples

To each ml of resuspended PEG concentrate, 3 ml of TRIzol LS (Invitrogen, Carlsbad, CA) and 800 µl of chloroform were added and norovirus RNA was extracted according the manufacturers' protocol. An additional purification tsep was conducted by adding 1 ml of 3M GuSCN lysis buffer to 1 ml of Trizol extracted RNA follwed by adding 2 ml of 100% EtOH., This mixture was then loaded onto RNAeasy midi columns (RNeasy Midi kit (Qiagen Inc, Valencia, CA). Each column was first washed with a mixture of GuSCN lysis buffer, Sigma water, and 100% EtOH (1:1:2). Second and third washes were preformed with 75% EtOH. The samples were then eluted with 225 µl 10 mM Tris-1mM EDTA (pH 8.0). The RNA was further concentrated by ethanol precipitation as follows: To 1 volume of RNA extract, 1/10 volume of 5M ammonium acetate, 1 µl of glycoblue, and 2 volumes of 100% ethanol were added and after incubation at -20°C for 30 min, the RNA was collected by centrifugation (13,000 x g, 20 min, 4°C). After a wash with 500 µl of 75% ethanol, the 'invisible' RNA pellet was resuspended in 10 µl of 10 mM Tris-HCl, 1 mM EDTA pH 8.0 and either tested directly or stored at -80 °C.

Conventional RT-PCR

The Qiagen One-Step RT-PCR kit™ (Qiagen Inc, Valencia, CA) without Q solution was used for all RT-PCR assays which were conducted in a final reaction volume of 25 µl. All oligonucleotide primers were used at 12.5 pmol per reaction. A total volume of 2.5 µl of viral RNA was reverse transcribed at 42°C for 30 min. followed by 15 min. at 95°C to activate the Taq polymerase. PCR consisted of 40 cycles at 94°C for 30 sec, 50°C for 30 seconds and 72°C for 30 sec.

Hemi-Nested PCR

Hemi-Nested RT-PCR was conducted in a final volume of 25 µl using the HotStarTaq Kit (Qiagen) with 1µl of template from the first round RT-PCR product. The PCR cycling conditions were as follows: 15 min at 95°C, followed by 40 cycles of 30 sec at 94°C, 30 sec at 50°C, and 1 min. at 72°C, with a final extension step of 10 min. at 72°C.

Norovirus TaqMan realtime RT-PCR

NoV TaqMan RT-PCR was carried out in 25 μl reactions containing 1μl of RNA template and reagents of the QuantiTect Probe RT-PCR kit (Qiagen). Separate monoplex reactions for GI and GII NoV were conducted in a Smartcycler (Cepheid, Sunnyvale, CA) as recently described (Jothikumar et al., 2005). Reaction conditions were as follows: reverse transcription for 30 min. at 50°C followed by activation of the Taq polymerase for 15 min. at 95°C, followed by 45 cycles of 10 sec at 95°C, 20 sec at 55°C, and 15 sec at 72°C. Amplification data was collected and analyzed with Smartcycler software version 2.0 (Cepheid).

Cloning of PCR products and DNA sequencing

PCR products selected for cloning were purified with the Qiagen QIAquick PCR purification Kit (Qiagen) and cloned into the TA TOPO kit (Invitrogen, Carlsbad, CA) strictly following the manufacturers' instructions using One Shot TOP10 chemically competent host cells. From each transformation, three white colonies were selected for DNA sequencing after overnight incubation in LB broth and plasmid purification using the Qiagen Plasmid Mini Kit (Qiagen). Plasmids were sequenced using standard M13 primers at the Lineberger Comprehensive Cancer Center Sequencing facility (UNC campus, Chapel Hill, NC) using the BigDye Terminator Cycle Sequencing Ready reaction Kit (PE Applied Biosystems). NoV sequences were analyzed using BioEdit sequence editor and compared with a region C local database based on a recently described norovirus typing system (Kageyama et al., 2004).

Reverse line blot hybridization of region B RT-PCR products

Broadly-reactive region B aminolinked GI and GII probes were covalently bound to an activated nylon membrane (Biodyne C; Pall Biosupport, East Hills, NY) using a miniblotter as described previously (Vinjé et al., 2000; Vinjé et al., 2004). Biotinylated region B RT-PCR products were generated using the GII primer set Mon431/Mon 433 as described previously (Vinjé et al., 2003) and RLB was performed at 48°C and specific binding to the probes was visualized using peroxidase labeled streptavidin followed by ECL detection on an X-ray film (Vinjé et al., 2000).

RESULTS

Selection of oligonucleotide hybride primer sets

We evaluated 13 different oligonucleotide primer combinations (7 GI and 6 GII) targeting the ORF1-ORF2 junction region and the 5'-region of ORF2 (region C) of human NoVs (Figure 4.1). Ten-fold serial dilutions of RNA from a panel of 7 GI strains (4 GI-4, 3 GI-2) and 8 GII strains (4 GII-4, and 4 GII-7 strains) (Table 4.2, strain panel #1) were analyzed. The One-Step RT-PCR products were analyzed after electrophoresis on a 2% agarose gel stained with ethidium bromide and the end-point titer of each primer pair for an individual NoV strain was determined as the highest dilution of viral RNA that showed a visible PCR product on gel (Figure 4.2).

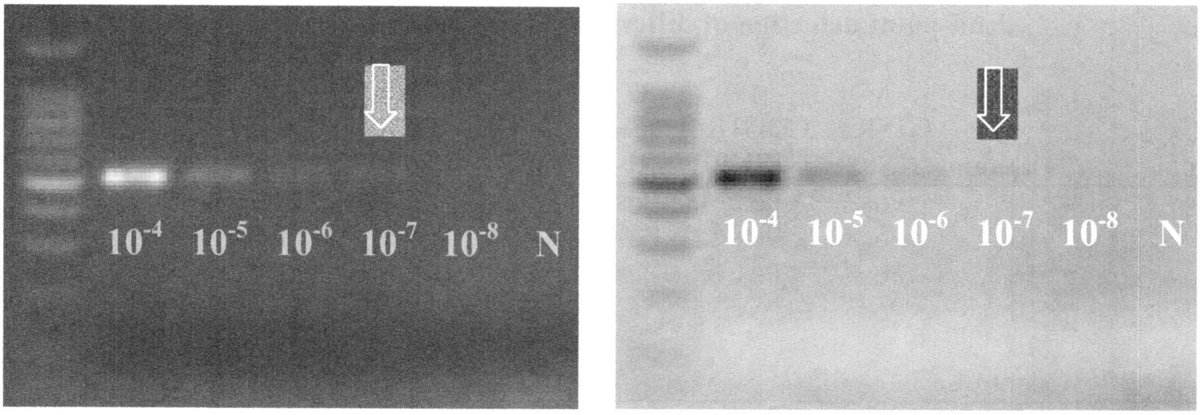

Figure 4.2 Detection limit estimation based on ten-fold serial dilutions using primers SRII-2 and SRII-1 with sample OB35

Table 4.3 shows a slight increase of the limit of detection of forward primer G1SKF compared to primer sets using the SRI-2 forward primer to detect 2 GI strains (Tables 4.4 and 4.5 contain complete data).

Table 4.3
Example of data obtained from the comparison of GI NoV primer combinations on a GI.2 strain (B39) and a GI.4 strain (S7)

			A	B	C	D	E	F	G
			G1SKF	COG1F	G1SKF	G1SKF	SRI-2	SRI-2	SRI-2
Sample	Dilution		G1SKR	G1SKR	SRI-1	SRI-3	SRI-1	G1SKR	SRI-3
B39	GI-4	10^{-4}	Pos	Pos	Pos	Pos	Pos	Pos	Pos
		10^{-5}	Pos	Pos	Pos	Pos	Pos	Pos	Pos
		10^{-6}	Pos	Pos	Pos	Pos	Pos	-	Pos
		10^{-7}	Pos	Pos	-	Pos	Pos	-	Pos
		10^{-8}	-	-	-	-	-	-	-
		control	-	-	-	-	-	-	-
S7	GI-2	10^{-4}	Pos	Pos	Pos	Pos	-	Pos	Pos
		10^{-5}	Pos	Pos	Pos	Pos	-	-	-
		10^{-6}	-	Pos	-	-	-	-	-
		10^{-7}	-	-	-	-	-	-	-
		control	-	-	-	-	-	-	-

Table 4.4
End-point detection of different GI primer combinations

Sample		Dilution	A G1SKF G1SKR	B COG1F G1SKR	C G1SKF SRI-1	D G1SKF SRI-3	E SRI-2 SRI-1	F SRI-2 G1SKR	G SRI-2 SRI-3
B39	GI-4	10^{-4}	Pos	Pos	Pos	Pos	Pos	Pos	Pos
		10^{-5}	Pos	Pos	Pos	Pos	Pos	Pos	Pos
		10^{-6}	Pos	Pos	Pos	Pos	Pos	-	Pos
		10^{-7}	Pos	Pos	-	Pos	Pos	-	Pos
		10^{-8}	-	-	-	-	-	-	-
		control	-	-	-	-	-	-	-
S11	GI-4	10^{-6}	Pos	Pos	Pos	Pos	-	Pos	Pos
		10^{-7}	Pos	Pos	Pos	Pos	-	Pos	Pos
		10^{-8}	Pos	-	Pos	Pos	-	-	-
		10^{-9}	Pos	-	Pos	Pos	-	-	-
		10^{-10}	-	-	Pos	-	-	-	-
		control	-	-	-	-	-	-	-
Sw1	GI-4	10^{-4}	Pos	Pos	Pos	Pos	Pos	Pos	Pos
		10^{-5}	Pos	Pos	Pos	Pos	-	Pos	Pos
		10^{-6}	Pos	-	Pos	Pos	-	-	Pos
		10^{-7}	Pos	-	Pos	Pos	-	-	-
		10^{-8}	-	-	-	-	-	-	-
		control	-	-	-	-	-	-	-
CDC4	GI-4	10^{-4}	Pos	Pos	Pos	Pos	-	Pos	Pos
		10^{-5}	Pos	Pos	Pos	Pos	-	Pos	Pos
		10^{-6}	Pos	Pos	pos	Pos	-	Pos	Pos
		10^{-7}	Pos	Pos	Pos	-	-	-	Pos
		10^{-8}	Pos	-	-	-	-	-	-
		control	-	-	-	-	-	-	-
Sw8	GI-2	10^{-4}	Pos	Pos	Pos	Pos	Pos	Pos	-
		10^{-5}	Pos	Pos	Pos	Pos	-	Pos	-
		10^{-6}	-	Pos	Pos	Pos	-	-	-
		10^{-7}	-	-	Pos	Pos	-	-	-
		10^{-8}	-	-	-	-	-	-	-
		control	-	-	-	-	-	-	-
CDC2	GI-2	10^{-2}	Pos	Pos	Pos	Pos	Pos	Pos	Pos
		10^{-3}	Pos	Pos	Pos	Pos	-	Pos	Pos
		10^{-4}	Pos	Pos	Pos	Pos	-	Pos	Pos
		10^{-5}	-	Pos	Pos	-	-	-	-
		10^{-6}	-	-	Pos	-	-	-	-
		control	-	-	-	-	-	-	-
S7	GI-2	10^{-4}	Pos	Pos	Pos	Pos	-	Pos	Pos
		10^{-5}	Pos	Pos	Pos	Pos	-	-	-
		10^{-6}	-	Pos	-	-	-	-	-
		control	-	-	-	-	-	-	-

Table 4.5
End-point detection of different GII primer combinations

Sample		Dilution	H G2SKF G2SKR	I JJG2Fw G2SKR	J JJG2Fw SRII-1	K SRII-2 SRII-3	L SRII-2 SRII-1	M SRII-2 G2SKR
OB35	GII-4	10^{-6}	Pos	Pos	-	Pos	Pos	Pos
		10^{-7}	Pos	-	-	-	Pos	-
		10^{-8}	Pos	-	-	-	Pos	-
		10^{-9}	-	-	-	-	Pos	-
		10^{-10}	-	-	-	-	-	-
		control	-	-	-	-	-	-
CDC11	GII-4	10^{-6}	Pos	Pos	-	-	Pos	Pos
		10^{-7}	Pos	Pos	-	-	Pos	-
		10^{-8}	Pos	Pos	-	-	Pos	-
		10^{-9}	-	-	-	-	Pos	-
		10^{-10}	-	-	-	-	-	-
		control	-	-	-	-	-	-
G1	GII-4	10^{-6}	Pos	Pos	-	-	Pos	-
		10^{-7}	Pos	Pos	-	-	-	-
		10^{-8}	-	-	-	-	-	-
		control	-	-	-	-	-	-
Sw24	GII-4	10^{-4}	Pos	Pos	-	-	-	Pos
		10^{-5}	Pos	Pos	-	-	-	Pos
		10^{-6}	Pos	Pos	-	-	-	-
		10^{-7}	Pos	Pos	-	-	-	-
		10^{-8}	-	Pos	-	-	-	-
		control	-	-	-	-	-	-
OB53	GII-7	10^{-2}	Pos	Pos	Pos	-	-	-
		10^{-3}	-	-	Pos	-	-	-
		10^{-4}	-	-	Pos	-	-	-
		10^{-5}	-	-	-	-	-	-
		control	-	-	-	-	-	-
J9a	GII-7	10^{-2}	Pos	Pos	Pos	-	-	-
		10^{-3}	Pos	Pos	-	-	-	-
		10^{-4}	Pos	-	-	-	-	-
		10^{-5}	-	-	-	-	-	-
		control	-	-	-	-	-	-
M13	GII-7	10^{0}	Pos	Pos	Pos	Pos	-	-
		10^{-1}	-	Pos	Pos	-	-	-
		10^{-2}	-	Pos	Pos	-	-	-
		10^{-3}	-	-	-	-	-	-
		control	-	-	-	-	-	-
CDC14	GII-7	10^{-2}	-	Pos	Pos	-	-	-
		10^{-3}	-	-	Pos	-	-	-
		10^{-4}	-	-	-	-	-	-
		control	-	-	-	-	-	-

Of all 7 different GI primer set combinations, 3 sets (primer set A; primer set B and primer set D) detected the highest dilutions of the 7 GI templates and two sets (primer set H and I) demonstrated the highest sensitivity for the panel of GII NoV strains.

To increase the detection limit we employed a second round of PCR, however; due to the tremendous sequence variation among GI and GII NoVs, only a few designated oligonucleotide primers internal to the first round primer sets could be identified and therefore we chose to evaluate the following hemi-nested assays:

For GI:

COG1F/G1SKR (primer set B) with G1SKF/ G1SKR (primer set A)
G1SKF/G1SKR (primer set A) with SRI-2/G1SKR (primer set F)

For GII:

COG2F/G2SKR (primer set I) with G2SKF/G2SKR (primer set H)

We compared the detection limit of these primer combinations with a hemi-nested region A primer set as well as a TaqMan real-time RT-PCR assay. For most strains, the addition of a nested round provided a 1 to 3 log increase in sensitivity for GI NoV and 1 to 5 log increase in sensitivity for GII NoV. However 50% of the GI strains tested did not show an increase in sensitivity with the addition of a nested round of PCR, rather the nested round showed equivalent sensitivity to single round PCR. Each region C hemi-nested primer set was 10 to 1000 times more sensitive than the region A hemi-nested primer set. Additionally, the TaqMan assays for both GI and GII norovirus showed comparable sensitivity to the hemi-nested primer set COG1F/G1SKR with G1SKF/G1SKR and COG2F/G2SKR with G2SKF/G2SKR, respectively (Tables 4.6 and 4.7) (Figures 4.3 & 4.4 and Figure 4.5).

Table 4.6
End-point detection of different GI NoV Hemi-nested primer combinations

			COG1F/G1SKR with G1SKF/G1SKR		G1SKF/G1SKR with SRI-2/G1SKF		MJV12/RegA with P290/RegA		TaqMan
Sample	Strain	Dilution	1°	2°	1°	2°	1°	2°	Ct Value
CDC 1	GI-1	10^{-1}	Pos	Pos	Pos	Pos	-	-	Pos
		10^{-2}	Pos	Pos	Pos	Pos	-	Pos	Pos
		10^{-3}	-	Pos	Pos	Pos	-	-	Pos
		10^{-4}	-	-	-	Pos	-	-	-
		10^{-5}	-	-	-	-	-	-	-
		Control	-	-	-	-	-	-	-
Sw 8	GI-2	10^{-5}	Pos	Pos	Pos	Pos	Pos	Pos	Pos
		10^{-6}	Pos	Pos	Pos	Pos	Pos	Pos	Pos
		10^{-7}	Pos	Pos	Pos	Pos	-	Pos	Pos
		10^{-8}	Pos	Pos	Pos	Pos	-	-	-
		10^{-9}	-	-	-	-	-	-	-
		Control	-	-	-	-	-	-	-
CDC 2	GI-2	10^{-3}	Pos	Pos	Pos	Pos	Pos	Pos	Pos
		10^{-4}	Pos	Pos	Pos	Pos	-	Pos	Pos
		10^{-5}	Pos	Pos	Pos	Pos	-	Pos	Pos
		10^{-6}	-	Pos	Pos	Pos	-	Pos	-
		10^{-7}	-	Pos	Pos	Pos	-	-	-
		10^{-8}	-	-	-	-	-	-	-
		Control	-	-	-	-	-	-	-
S7	GI-2	10^{-3}	Pos	Pos	Pos	Pos	-	Pos	Pos
		10^{-4}	Pos	Pos	Pos	Pos	-	Pos	Pos
		10^{-5}	Pos	Pos	Pos	Pos	-	Pos	Pos
		10^{-6}	Pos	Pos	-	-	-	-	-
		10^{-7}	-	-	-	-	-	-	-
		Control	-	-	-	-	-	-	-

(continued)

Table 4.6 (Continued)
End-point detection of different GI NoV hemi-nested primer combinations

Sample	Strain	Dilution	COG1F/G1SKR with G1SKF/G1SKR		G1SKF/G1SKR with SRI-2/G1SKF		MJV12/RegA with P290/RegA		TaqMan
			1°	2°	1°	2°	1°	2°	Ct Value
Sw 6	GI-3	10^{-3}	Pos	Pos	Pos	Pos	-	Pos	Pos
		10^{-4}	Pos	Pos	Pos	Pos	-	Pos	Pos
		10^{-5}	Pos	Pos	Pos	Pos	-	Pos	Pos
		10^{-6}	Pos	Pos	-	Pos	-	Pos	-
		10^{-7}	-	-	-	-	-	-	-
		Control	-	-	-	-	-	-	-
Sw 18	GI-3	10^{-5}	Pos	Pos	Pos	Pos	-	-	Pos
		10^{-6}	Pos	Pos	-	-	-	-	Pos
		10^{-7}	Pos	Pos	-	Pos	-	-	Pos
		10^{-8}	-	-	-	-	-	-	-
		Control	-	-	-	-	-	-	-
B39	GI-4	10^{-4}	-	Pos	Pos	Pos	-	Pos	Pos
		10^{-5}	Pos	Pos	Pos	Pos	-	Pos	Pos
		10^{-6}	-	-	Pos	Pos	-	-	-
		10^{-7}	Pos	Pos	-	-	-	-	-
		10^{-8}	-	-	-	-	-	-	-
		Control	-	-	-	-	-	-	-
Sw 1	GI-4	10^{-1}	Pos	Pos	Pos	Pos	-	Pos	Pos
		10^{-2}	Pos	Pos	Pos	Pos	-	Pos	Pos
		10^{-3}	Pos	Pos	Pos	Pos	-	Pos	Pos
		10^{-4}	Pos	Pos	Pos	Pos	-	Pos	Pos
		10^{-5}	Pos	Pos	Pos	Pos	-	-	Pos
		10^{-6}	-	-	-	-	-	-	-
		Control	-	-	-	-	-	-	-

(continued)

Table 4.6 (Continued)
End-point detection of different GI NoV hemi-nested primer combinations

Sample	Strain	Dilution	COG1F/G1SKR with G1SKF/G1SKR 1°	COG1F/G1SKR with G1SKF/G1SKR 2°	G1SKF/G1SKR with SRI-2/G1SKF 1°	G1SKF/G1SKR with SRI-2/G1SKF 2°	MJV12/RegA with P290/RegA 1°	MJV12/RegA with P290/RegA 2°	TaqMan Ct Value
CDC 4	GI-4	10^{-5}	Pos	Pos	Pos	Pos	-	Pos	Pos
		10^{-6}	Pos	Pos	Pos	Pos	-	Pos	Pos
		10^{-7}	Pos	Pos	Pos	Pos	-	-	Pos
		10^{-8}	-	-	-	-	-	-	-
		Control	-	-	-	-	-	-	-
S11	GI-4	10^{-5}	Pos	Pos	Pos	Pos	-	Pos	n/a
		10^{-6}	Pos	Pos	Pos	Pos	-	Pos	n/a
		10^{-7}	Pos	Pos	Pos	Pos	-	-	n/a
		10^{-8}	-	-	-	-	-	-	n/a
		Control	-	-	-	Pos	-	-	n/a
B49	GI-6	10^{-1}	Pos	Pos	Pos	Pos	Pos	Pos	Pos
		10^{-2}	Pos	Pos	Pos	Pos	Pos	Pos	Pos
		10^{-3}	-	Pos	Pos	Pos	-	Pos	Pos
		10^{-4}	-	Pos	Pos	Pos	-	Pos	-
		10^{-5}	-	Pos	-	-	-	Pos	-
		10^{-6}	-	-	-	-	-	-	-
		Control	-	-	-	-	-	-	-
Sw 3	GI-6	10^{-1}	-	-	Pos	Pos	-	-	Pos
		10^{-2}	-	-	-	-	-	-	Pos
		10^{-3}	-	-	-	-	-	-	Pos
		Control	-	-	-	-	-	-	-

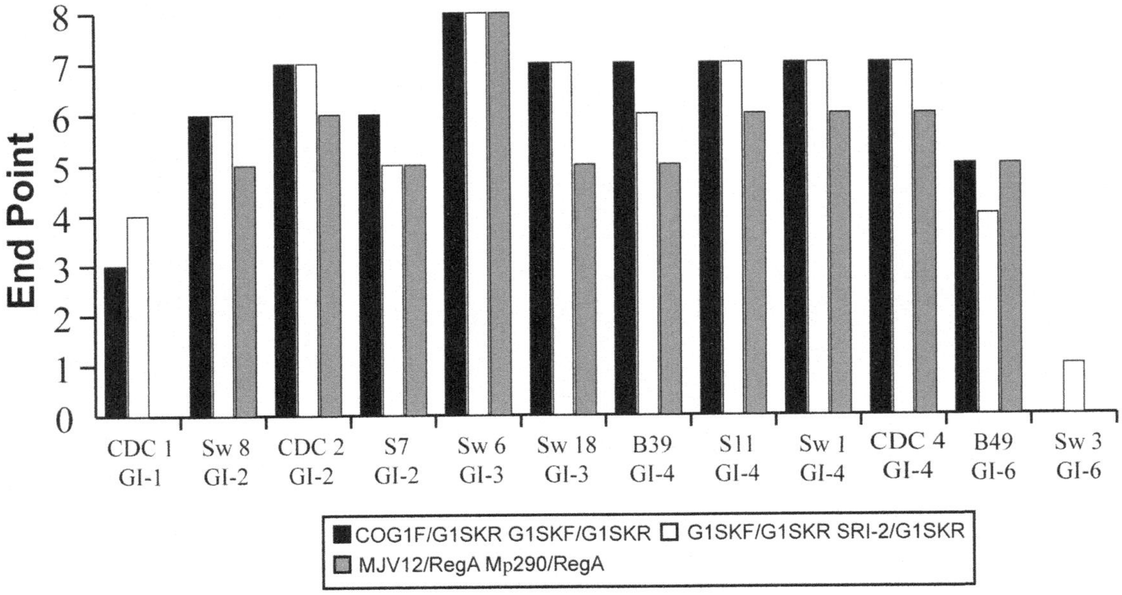

Figure 4.3 Comparison of hemi-nested assays for the detection of GI NoV

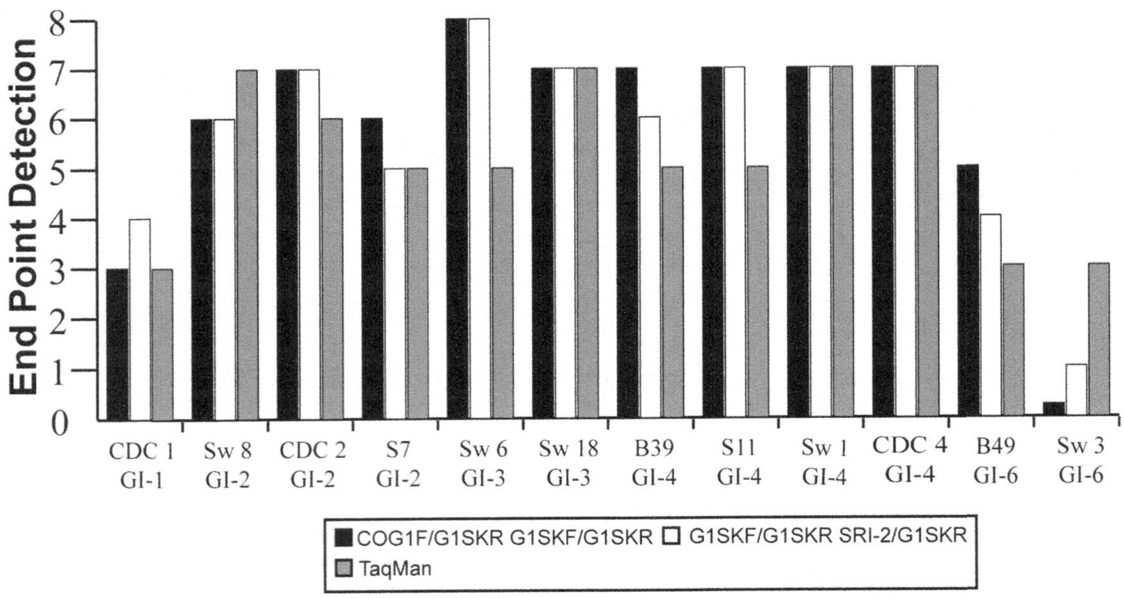

Figure 4.4 Comparison of hemi-nested and TaqMan assays for the detection of GI No

Table 4.7
End-point detection of different GII NoV hemi-nested primer combinations

Sample	Strain	Dilution	COG2F/G2SKR with G2SKF/G2SKR 1°	2°	MJV12/RegA with Mp290/RegA 1°	2°	TaqMan Ct Value
Sw 7	GII-1	10^{-3}	Pos	Pos	-	Pos	28.87
		10^{-4}	-	Pos	-	Pos	32.56
		10^{-5}	-	Pos	-	Pos	36.56
		10^{-6}	-	Pos	-	Pos	39.37
		10^{-7}	-	Pos	-	Pos	-
		10^{-8}	-	-	-	-	-
		Control	-	-	-	-	-
B13	GII-2	10^{-1}	-	Pos	Pos	Pos	38.67
		10^{-2}	-	Pos	-	Pos	31.57
		10^{-3}	-	Pos	-	Pos	-
		10^{-4}	-	Pos	-	Pos	-
		10^{-5}	-	Pos	-	Pos	-
		10^{-6}	-	Pos	-	-	-
		Control	-	-	-	-	-
B40	GII-3	10^{-4}	Pos	Pos	-	Pos	33.72
		10^{-5}	Pos	Pos	-	Pos	37.12
		10^{-6}	-	Pos	-	Pos	-
		10^{-7}	-	-	-	-	-
		Control	-	-	-	-	-
OB35	GII-4	10^{-2}	Pos	Pos	Pos	Pos	29.59
		10^{-3}	Pos	-	Pos	Pos	34.06
		10^{-4}	-	-	Pos	Pos	37.52
		10^{-5}	-	-	Pos	Pos	-
		10^{-6}	-	-	Pos	-	42.92
		10^{-7}	-	-	-	-	-
		Control	-	-	-	-	-
G1	GII-4	10^{-3}	Pos	-	Pos	Pos	31.92
		10^{-4}	Pos	-	Pos	Pos	34.40
		10^{-5}	-	-	Pos	Pos	38.53
		10^{-6}	-	-	-	-	-
		10^{-7}	-	-	-	Pos	-
		10^{-8}	-	-	-	-	-
		Control	-	-	-	-	-
Sw 24	GII-4	10^{-3}	-	Pos	-	-	39.28
		10^{-4}	-	Pos	-	-	42.11
		10^{-5}	-	-	-	-	-
		Control	-	-	-	-	-
CDC11	GII-4	10^{-1}	Pos	Pos	Pos	Pos	27.67
		10^{-2}	Pos	Pos	Pos	Pos	32.47
		10^{-3}	Pos	Pos	Pos	Pos	39.18
		10^{-4}	-	Pos	-	Pos	42.99
		10^{-5}	-	Pos	-	Pos	-
		10^{-6}	-	Pos	-	Pos	-
		Control	-	-	-	-	-
Sw 9	GII-5	10^{-1}	-	Pos	-	Pos	37.94
		10^{-2}	-	-	-	-	-
		Control	-	-	-	-	-

(continued)

Table 4.7 (continued)
End-point detection of different GII NoV hemi-nested primer combinations

Sample	Strain	Dilution	COG2F/G2SKR with G2SKF/G2SKR 1°	COG2F/G2SKR with G2SKF/G2SKR 2°	MJV12/RegA with Mp290/RegA 1°	MJV12/RegA with Mp290/RegA 2°	TaqMan Ct Value
Sw 11	GII-6	10^{-2}	Pos	Pos	-	Pos	30.72
		10^{-3}	Pos	Pos	-	-	33.76
		10^{-4}	-	Pos	-	-	38.83
		10^{-5}	-	Pos	-	-	-
		10^{-6}	-	-	-	-	-
		Control	-	-	-	-	-
CDC14	GII-7	10^{-1}	Pos	Pos	-	Pos	28.11
		10^{-2}	-	Pos	-	Pos	32.20
		10^{-3}	-	Pos	-	Pos	37.85
		10^{-4}	-	Pos	-	-	39.68
		10^{-5}	-	-	-	-	-
		Control	-	-	-	-	-
M13	GII-7	10^{-1}	Pos	Pos	-	Pos	32.04
		10^{-2}	Pos	Pos	-	Pos	34.56
		10^{-3}	-	Pos	-	-	38.74
		10^{-4}	-	-	-	-	42.80
		10^{-5}	-	-	-	-	-
		Control	-	-	-	-	-
OB 53	GII-7	10^{-1}	-	-	Pos	Pos	22.49
		10^{-2}	-	Pos	Pos	Pos	26.65
		10^{-3}	-	Pos	-	Pos	31.61
		10^{-4}	-	-	-	Pos	36.77
		10^{-5}	-	-	-	-	39.33
		10^{-6}	-	-	-	-	42.72
		Control	-	-	-	-	-

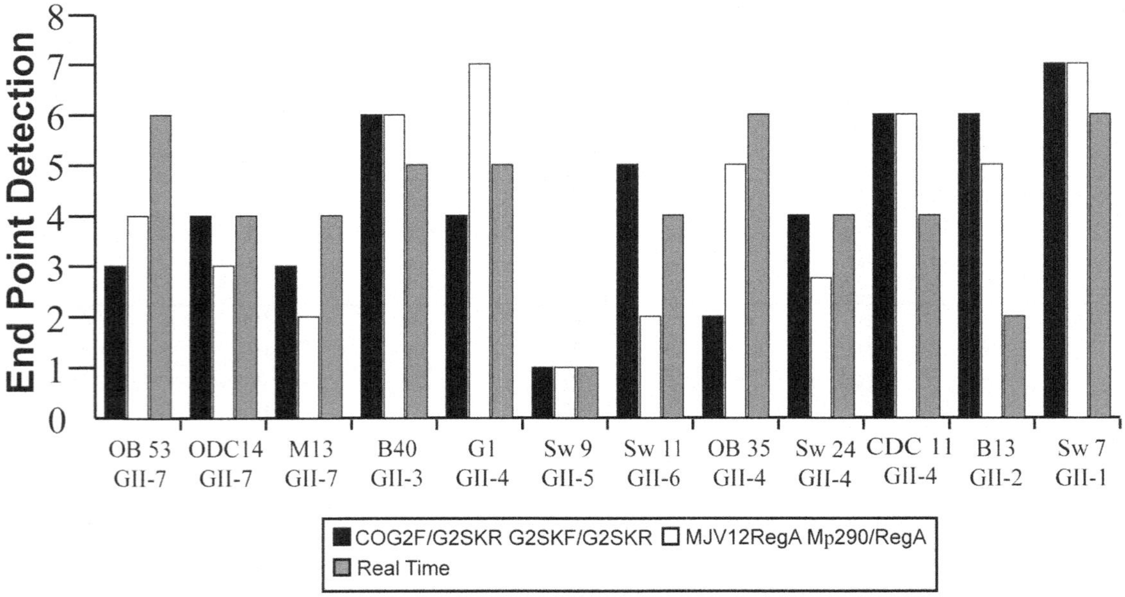

Figure 4.5 Comparison of hemi-nested and TaqMan assays for the detection of GII NoV

Environmental Samples -Waste water concentrates

Wastewater concentrates were analyzed by conventional hemi-nested RT-PCR using GI primer sets COG1F/G1SKR (primer set B) and G1SKF/G1SKR (primer set A) and for GII using COG2F/G2SKR (primer set I) and G2SKF/G2SKR (primer set H). For GI NoV, RT-PCR products of the appropriate size were observed for 47.8% (11/23) of the influent and 82.6% (19/23) of the effluent samples (Figure 4.6). For GII norovirus, multiple bands were visible on the gel making identifying clear potential positives difficult (Figure 4.7).

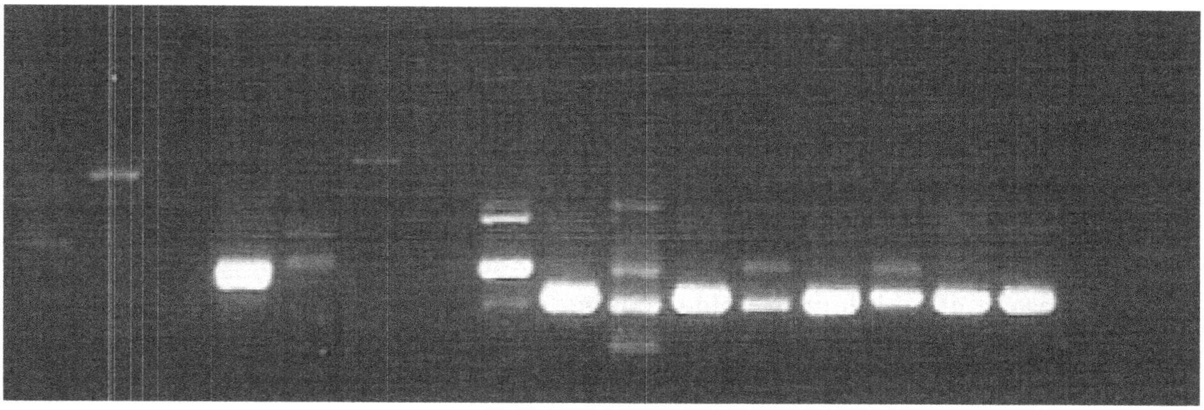

Figure 4.6 Results of ethidium-bromide staining of GI hemi nested PCR primer set B products (329 bp) on a 2% agarose gel. The templates were RNA from wastewater concentrates.

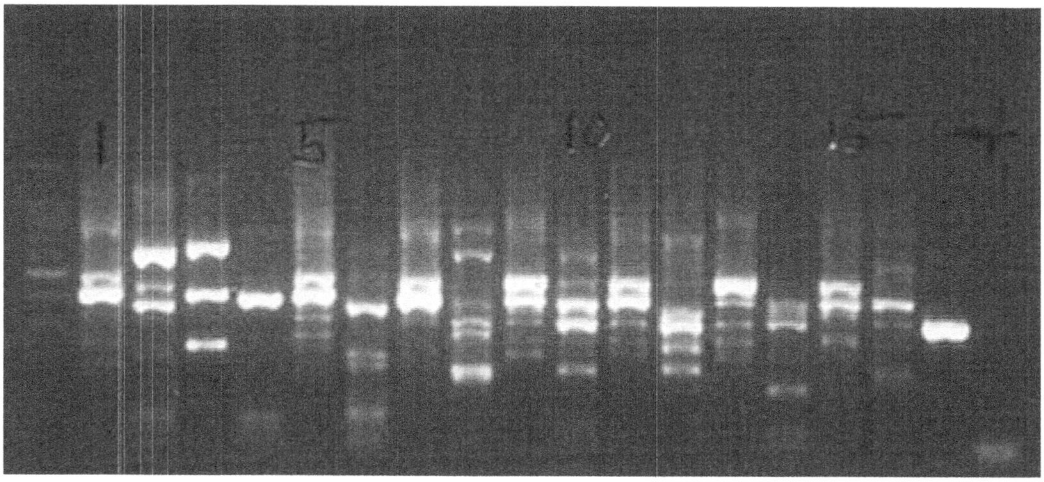

Figure 4.7 Results of ethidium-bromide staining of GII hemi nested PCR primer set H products (343 bp) on a 2% agarose gel. The templates were RNA from wastewater concentrates.

Therefore, we retested the RNA samples with primer set Mon431 / Mon433, targeting region B of the norovirus genome. Although the agarose gel results of this assay still showed a number of multiple bands (Figure 4.8), bands of expected size were identifiable for most samples. However Southern hybridization or sequencing was not performed to confirm that the products having the appropriate size (213 bp) were indeed NoV sequences.

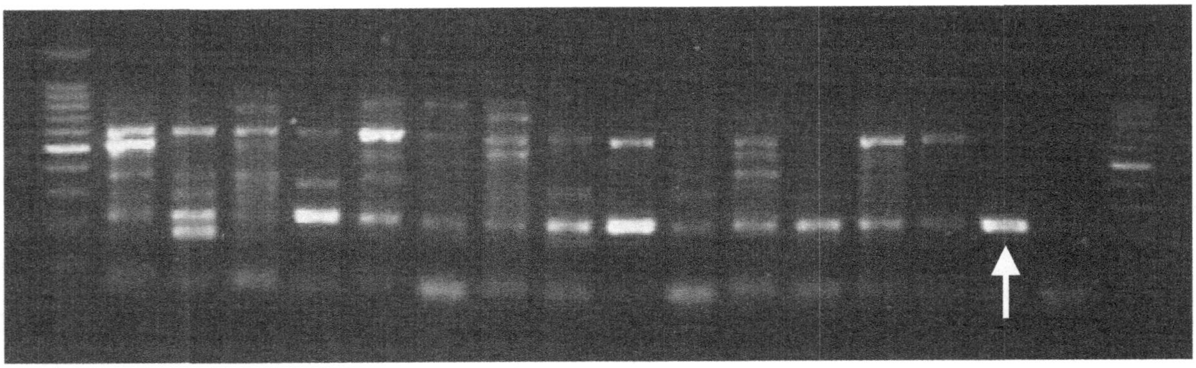

Figure 4.8 Results of ethidium-bromide staining of region B GII RT-PCR products (213 bp) on a 2% agarose gel. The arrow indicates the GII positive control.

We tested RNA from the 46 wastewater samples with a recently developed norovirus real-time TaqMan RT-PCR assays for GI and GII norovirus (see chapter 3). In total, 21 (45.6%) of the 46 samples tested positive for GI with 3 (13%) of the 23 influent and 18 (78%) of the 23 effluent samples testing positive. GII norovirus RNA was detected in 31 (67.3%) of the 46 samples with 16 (69.5%) and 15 (65.2%) of the influent and effluent samples testing positive, respectively (see Table 4.8, Figure 4.9, and Figure 4.10).

Table 4.8

Comparison of TaqMan and conventional RT-PCR for the detection of GI and GII Norovirus RNA from wastewater sample concentrates.

			GI Norovirus - Waste Water Concentrates			GII Norovirus - Waste Water Concentrates		
Month	Facility		COG1F/G1SKR with G1SKF/G1SKR	TaqMan Ct Value	Cloning with Sequencing	COG2F/G2SKR with G2SKF/G2SKR	TaqMan Ct Value	Region B GII PCR-RLB
Sept	Cobb Co.	Influent	-	38.8		-	34.77	+
	Cobb Co.	Effluent	-	35.0		-	36.88	-
	R.M. Clayton	Influent	+	-		++	42.38	-
	R.M. Clayton	Effluent	-	-		+	-	-
Aug	Cobb Co.	Influent	-	34.8		++	-	-
	Cobb Co.	Effluent	-	-		-	-	-
	R.M. Clayton	Influent	+	-		-	-	-
	R.M. Clayton	Effluent	+++	39.9	Sequenced GI w/o cloning	-	-	-
July	Cobb Co.	Influent	++	-		+	-	-
	Cobb Co.	Effluent	+++	35.3	GI-1 and GI-2	-	-	-
	R.M. Clayton	Influent	+	-		+	-	-
	R.M. Clayton	Effluent	+++	-	GI-7	+	41.20	-
June	Cobb Co.	Influent	++	-		-	39.93	-
	Cobb Co.	Effluent	+++	30.1	GI-1	+	31.39	+
	R.M. Clayton	Influent	-	-		-	-	-
	R.M. Clayton	Effluent	+++	36.4	GI-2	-	-	-
May	Cobb Co.	Influent	-	-		-	38.64	-
	Cobb Co.	Effluent	+++	30.0	GI-2	-	31.36	+
	R.M. Clayton	Influent	-	-		-	-	-
	R.M. Clayton	Effluent	++	-	GI-1	-	-	-
Apr	Cobb Co.	Influent	+++	-	GI-2	-	34.74	-
	Cobb Co.	Effluent	+++	31.9	GI-2	+	32.95	+
	R.M. Clayton	Influent	-	-		n/a	37.32	-
	R.M. Clayton	Effluent	+++	30.7	GI-2	n/a	30.84	+
Mar	Cobb Co.	Influent	++	-	GI-1	+	30.44	+
	Cobb Co.	Effluent	+++	34.2	GI-1	-	32.27	+
	R.M. Clayton	Influent	+++	40.2		n/a	33.07	-
	R.M. Clayton	Effluent	+++	32.1		n/a	31.31	+

(continued)

Table 4.8 (continued)
Comparison of TaqMan and conventional RT-PCR for the detection of GI and GII Norovirus RNA from wastewater sample concentrates.

			GI Norovirus - Waste Water Concentrates			GII Norovirus - Waste Water Concentrates		
Month	Facility		COG1F/G1SKR with G1SKF/G1SKR	TaqMan Ct Value	Cloning with Sequencing	COG2F/G2SKR with G2SKF/G2SKR	TaqMan Ct Value	Region B GII PCR-RLB
Feb	Cobb Co.	Influent	+++	-		+	26.94	-
	Cobb Co.	Effluent	++	34.3		-	32.57	+
	R.M. Clayton	Influent	+++	-		n/a	30.37	-
	R.M. Clayton	Effluent	+++	31.7		n/a	30.59	+
Jan	Cobb Co.	Influent	-	-		-	40.80	-
	Cobb Co.	Effluent	+++	35.7		-	43.85	+
	R.M. Clayton	Influent	+++	-		n/a	33.88	-
	R.M. Clayton	Effluent	+++	29.6		n/a	33.26	-
Dec	Cobb Co.	Influent	-	-		+	39.34	-
	Cobb Co.	Effluent	+++	38.0		-	42.41	+
	R.M. Clayton	Influent	-	-		n/a	37.82	-
	R.M. Clayton	Effluent	++	42.5		n/a	34.14	-
Nov	Cobb Co.	Influent	-	-		+	34.15	-
	Cobb Co.	Effluent	++	39.9		-	39.22	-
	R.M. Clayton	Influent	-	-		n/a	40.93	-
	R.M. Clayton	Effluent	-	38.2		n/a	-	-
Oct	Cobb Co.	Influent	-	-		n/a	-	-
	Cobb Co.	Effluent	++	-		n/a	-	-

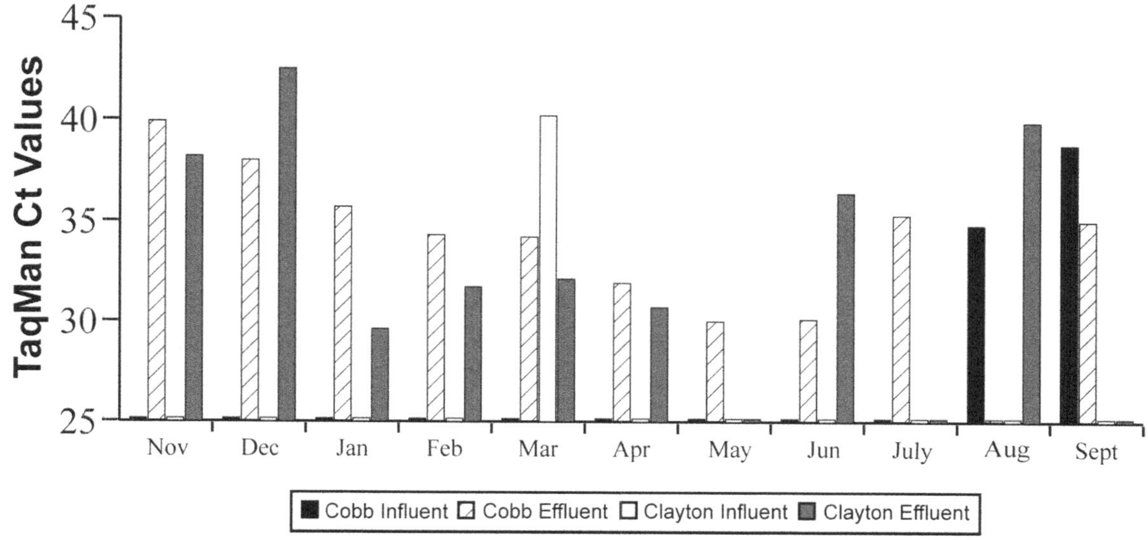

Figure 4.9 Seasonal distribution of GI NoVs in waste water concentrates

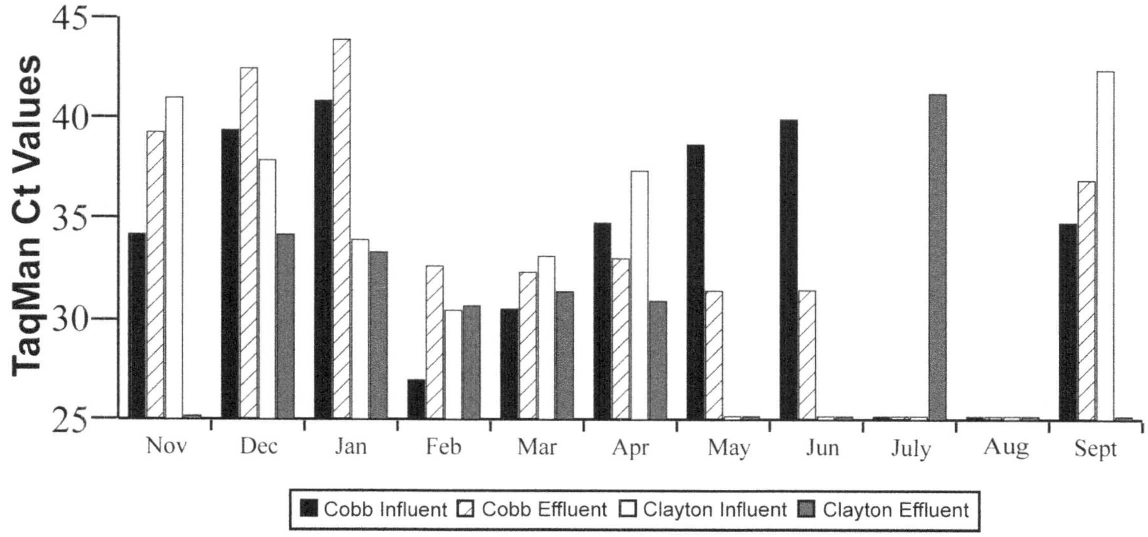

Figure 4.10 Seasonal distribution of GII NoVs in waste water concentrates

Drinking water concentrates

Drinking water samples were batch tested (three samples per group) for GI noroviruses by conventional hemi-nested RT-PCR with primer set COG1F/G1SKR with G1SKF/G1SKR and no positives were detected. Similarly, none of the samples tested positive for GI or GII norovirus by real-time TaqMan RT-PCR.

RT-PCR Inhibitor screening

To test for false negatives due to the presence of RT-PCR inhibitors, we mixed RNA from five wastewater samples (effluent) that tested negative by TaqMan RT-PCR with a known amount of GII RNA. The Ct values from all five samples did not differ compared the spiked Sigma water sample (Table 4.9).

Table 4.9
Analyzing water concentrates for the presence of RT-PCR inhibitors using a known amount of GII.4 RNA by TaqMan RT-PCR

Sample	Ct WW sample	Ct WW sample GII.4	Ct GII.4
June A WW N	neg	31.88	31.82
May A WW N	neg	31.47	30.82
Aug C WW N	neg	31.86	30.74
Sept A WW F	neg	31.23	30.88
July A WW N	neg	30.21	30.78

Cloning

RT-PCR hemi-nested PCR products of the appropriate size of 11 sewage samples were cloned and sequenced. From 9 samples only one clone was sequenced to confirm the presence of NoV. Of 2 samples, 3 clones were sequenced to identify possible multiple GI genotypes. All sequences were compared to a local BLAST database of GI region C NoV to identify the closest GI genotype strain and all sequences could be easily grouped into GI-1, GI-2, and GI-7 genotypes. Of the two samples from which multiple colonies were purified and sequenced, both contained multiple strains (Table 4.8).

Reverse Line Blot Hybridization

All 46 samples assayed by conventional region B GII RT-PCR with primer set Mon431/Mon433 were assayed by reverse line blot hybridization. In total, 12 (26%) of the 46 samples tested positive for GII norovirus. Of these, 2 (8%) of the 23 were influent samples and 10 (43%) of the 23 were effluent samples (data not shown).

DISCUSSION

Noroviruses (NoVs) are recognized as the leading cause of acute outbreaks and cases of nonbacterial gastroenteritis in humans impacting millions of people each year worldwide.
Recent RT-PCR assays have been shown to be highly successful for the detection of NoVs in clinical samples such as stool samples and emesis. However, since the detection limit for each NoV genotype can vary significantly (Vinjé et al., 2003), little data are available on the detection of low copy numbers of NoVs which can be expected in environmental samples.

In this study, several different hybrid primer combinations targeting the 3'-end of ORF1 and the 5'-end of ORF2 were evaluated for their detection limit using a panel of different NoV GI and GII genotypes. As the most commonly documented vehicles for infection include municipal drinking water, well water, and recreational waters, a selection of the primers sets that showed the highest detection limit (i.e., lowest detectable copy number) were further evaluated using naturally contaminated environmental samples obtained from sewage and drinking water facilities in Georgia. Detection of low copy numbers in large volumes of water that may contain humic and fulvic acid components, which are considered potent PCR inhibitors, have made the development of a broadly-reactive sensitive assay for the detection of NoVs in environmental samples a real challenge.

Not surprisingly, this study identified a high number of the waste water samples as positive for norovirus. Based on the TaqMan RT-PCR assay, NoV positive rates were 45% for GI norovirus and 65% for GII norovirus. Positive rates for NoV can be as high as 100% in domestic sewage influent and 89% in treated wastewater effluent samples (Ueki et a 2005). Based on the TaqMan assays, the concentration of norovirus found in this study ranged from 10^2 to 10^7 PCR detectable units (PDU) for raw sewage and 10^2 to 10^6 PDU for treated sewage. These results are similar to previously reported concentrations of norovirus ranging from 10^4 to 10^6 PDU per liter in raw sewage and 10^3 to 10^4 pdu/l in treated sewage (van den Berg et al., 2005; Lodder et al., 2005).

Surprisingly, almost half of the sample sets (one influent and one effluent from the same facility) had influent samples that were negative for norovirus by TaqMan RT-PCR while the effluents tested positive. These unusual results contradict results from other studies where a typical reduction of virus RNA between influent and effluent ranging between 0.7 to 2.7 log10 units was reported (Lodder et al., 2005; van den Berg et al., 2005).

The sensitivity of an RT-PCR assay is a careful balance between the design of the assay (including the selection and concentrations of primer and probes) and the quality of the sample. Variation between samples includes both the recovery efficiency of the viral extraction method used and the ability of the method to remove inhibitors (Ando et al., 1995; Hale et al., 1996). Because each of the waste water samples was concentrated and extracted following the same method, the variability in the recovery of RNA is limited. A rational explanation would be that the influent sewage samples contain greater amounts of inhibitory substances which are interfering with the sensitivity of the assay. Previous studies have found up to a 75% reduction in the recovery of RNA from spiked water samples attributable to inhibitory substances present in sewage samples (Laverik et al., 2004). The amount of inhibitors is expected to be higher in influent samples of sewage due to its high organic content. This could partly explain the low positive rate in sewage influent samples.

With treatment plants discharging enteric viruses at concentrations as high as 10^4 pdu/l into surface waters, combined with natural sources from runoff, it is certainly possible to find norovirus in drinking water intake (Lodder et al., 2005). Norovirus has even been detected in up to 10.2% of tap water samples in Japan (Haramoto et al., 2004). In this study, none of the drinking water samples, neither influent nor effluent, tested positive for norovirus, but due to the size of the sample this is not surprising. However, only a small portion of the sample concentrates was tested. For influent samples, only 2% (an equivalent of 100 ml) of the total sample was tested by each NoV TaqMan assay and 5% (equivalent to 250 ml) in a hemi-nested PCR assay. For effluent samples, 2% (equivalent to 2000 ml) of the total sample was tested in each TaqMan reaction and 5% (equivalent to 5000 ml) was tested in each hemi-nested assay. These volumes are relatively low compared to other reported studies where NoV was detected in drinking water. Haramoto and co-workers, for example tested concentrated samples equivalent to a range of 5.5 to 27.5 liters of drinking water (Haramoto et al., 2004). Therefore, in future studies further concentrating te water samples would be recommended which likely will increase the amount of NoV positives.

Additionally, since drinking water samples are expected to contain lower copy numbers the co-concentration of RT-PCR inhibitors, perhaps present in beef extract used for eluting the viruses from the 1-MDS filters might account for samples testing false negative (Haramoto et al., 2004; Wyn-Jones et al., 2000).

Most of the samples that tested negative for norovirus and were screened for the presence of inhibitors tested negative Perhaps, lowering the number of viral copies spiked into each sample to screen for inhibitors would more accurately test for the inhibition of low copy numbers. Since GII NoV strains are consistently found to be more prevalent (>90%) than GI strains, it is likely that higher concentrations of GII norovirus are also present in sewage contaminated waters.

Due to the co-circulation of genetically different norovirus strains in the community at one give time, it is common to detect multiple strains of norovirus in sewage and surface water samples (van den Berg et al., 2005) with as many as 10 different identified within a single year (Lodder et al., 2005). In our study, we were also able to detect several genetically different strains in 1 water concentrate. This genetic diversity of strains further highlights how critical it is to use well-evaluated broadly reactive primer sets that amplify different strains with similar efficiency.

Hemi-nested RT-PCR, while more time consuming than TaqMan RT-PCR, remains the method of choice for the source tracking of outbreak samples since the small unique sequences (~30 nt) that potentially can be obtained from realtime TaqMan RT-PCR assays does not allow to reliably link virus detected in contaminated water samples with clinical cases. However, unlike TaqMan RT-PCR, hemi-nested products require the need for confirmation by hybridization or sequencing of the RT-PCR products (Vinjé et al., 2003). This is especially true for environmental samples which often generate multiple bands on gel when visualized by UV illumination. The TaqMan assay allows simultaneous confirmation of the PCR products which removes the time and cost of additional confirmation assays needed for conventional RT-PCR assays. The speed and sensitivity of TaqMan RT-PCR assays make this an ideal method for the rapid detection of low copies of norovirus in environmental samples. Recent reports using TaqMan-based RT-PCR assays for NoV demonstrated that these assays can be used reliably for rapid and sensitive detection of low copy numbers in environmental matrices such as shellfish (Nishida et al., 2003; Jothikumar et al., 2005).

A variation to the nested PCR has recently been reported. This so-called booster RT-PCR, which uses two rounds of RT-PCR with the same primer set in each round, has been proposed as a more sensitive detection method compared to hemi-nested RT-PCR (De Medici et al., 2004). However this method also has a significant disadvantage since non-specific binding occurring during the first round would be re-amplified in the second round, thus reducing the specificity of the assay. The loss in specificity would increase the number of false positives and increase the number of samples needing confirmatory sequencing or hybridization thus raising the costs and time required.

In our study, it appears that for the detection of GI NoVs, the hemi-nested RT-PCR (65% positive rate) is better able to detect GI NoV than TaqMan RT-PCR (45% positive rate). Of the 46 samples tested for GI NoV, 13 samples were potentially positive for GI norovirus and negative by TaqMan RT-PCR. Four samples that tested negative by TaqMan RT-PCR were positive for GI NoV by conventional hemi-nested PCR after cloning and sequencing demonstrating the further fine-tuning of the TaqMan assays is required. Recent data show that the inclusion of an additional GI-probe significantly improves the specificity of the GI-assay (J. Vinjé, personal communication).

However, for the detection of GII NoVs, the TaqMan assay (67% positive rate) is superior compared to the hemi-nested RT-PCR assays surveyed. The COG2F/G2SKR with G2SKF/G2SKR set was unable to clearly define positive results due to the multiple banding of products. The alternative primer set Mon433/Mon431 combined with a confirmatory RLB step (26% positive rate) was also clearly inferior to the GII TaqMan assay. Clearly, TaqMan RT-PCR is more reliable and sensitive with the detection of GII NoV from environmental samples. However, for source tracking contaminated (drinking) water or shellfish to an outbreak, sequencing information is required and further research into a conventional NoV RT-PCR assay is highly recommended.

In conclusion, detection of NoVs in environmental sample concentrates may require multiple approaches due to the differences of sensitivity in the detection of different genogroups among available assays. Despite the need for additional confirmatory steps, the hemi-nested RT-PCR assay using the primer set COG1F/G1SKR with G1SKF/G1SKR was compared to the real-time TaqMan RT-Assay more sensitive to detect GI noroviruses in environmental sample concentrates. Conversely, TaqMan RT-PCR was far superior in the detection of GII NoV compared to hemi-nested RT-PCR assay using the primer set COG2F/G2SKR with G2SKF/G2SKR.

CHAPTER 5
SUMMARY AND CONCLUSIONS

Widespread use of molecular methods for the detection of noroviruses (NoVs) in environmental water samples has been hampered by the unavailability of sensitive and robust methods to detect this group of genetically diverse viruses at low copy numbers. Currently, several broadly reactive RT-PCR assays targeting different regions of the viral genome have been described; however, these assays are primarily optimized for the detection of NoVs in fecal specimens with a relatively high viral load. In this study we developed and evaluated novel methods for the concentration, detection and genotyping of NoVs from complex environmental matrices. Specific conclusions of the project are:

1. As a proof-of-concept, we developed a novel assay based on H-type 1 histo-bloodgroup carbohydrate magnetic beads (HGBA-beads) which had a detection limit of 20-200 RT-PCR detectable units of GI.1 norovirus.

2. GI.1 NoV particles bound to the HGBA-beads were resistant to RNase treatment indicating the protein capsid of the bound viruses is intact and thus potentially infectious.

3. This study provides the framework for implementing CMS for concentration and recovery of low copy numbers of NoV from environmental waters for detection by molecular methods. We hypothesize that binding of human NoV to HBGAs followed by detection by RT-PCR may be a surrogate for detecting infectious viruses.

4. In the present study, we have developed A rapid (less than 90 min.) and broadly reactive one-step TaqMan® RT-PCR assays for the detection of GI and GII NoV in fecal samples as well as shellfish samples. The specificity and sensitivity of all steps of the assays were systematically evaluated and in its final format, the monoplex assays were validated using RNA extracted from a panel of 84 stool specimens including NoV strains representing 19 (7 GI , 11 GII, and 1 GIV) different genotypes. The assays were further validated on 38 shellfish cDNA extracts previously tested by nested PCR.

5. Comparison with a recently published real-time assay showed that our one-step TaqMan® RT-PCR assays had a significantly higher sensitivity, and was at least as sensitive as nested PCR. For stool specimens, a duplex TaqMan® RT-PCR assay using GI and GII TaqMan® probes labeled with FAM and JOE, respectively, performed equally well compared to individual genogroup-specific monoplex assays. All other tested enteric viruses tested negative and no cross-reaction between genogroups was observed.

6. In this study, we compared the detection limit of 13 (7 GI, 6 GII) different norovirus oligonucleotide primer combinations targeting the 5'-end of ORF2 (region C) by conventional RT-PCR followed by hemi-nested PCR. A panel of 24 norovirus strains including four different GI genotypes (GI.1, GI.2, GI.3, GI.4) and seven different GII genotypes (GII.1 through GII.7) was used to evaluate the different primer combinations of which COG1F/G1SKR and G1SKF/G1SKR for GI and COG2F/G2SKR and

G2SKF/G2SKR for GII NoV showed the lowest detection limit across the range of genotypes tested.

7. We compared these hemi-nested PCR assays with a recently developed TaqMan®-based real-time RT-PCR assay (see previous chapter) on naturally contaminated water samples collected from Sept 2003 –Sept 2004. None of the drinking water samples and 63% of the waste water samples tested positive for NoVs. While hemi-nested PCR is as sensitive as real-time PCR detection and is capable to compare clinical and environmental strains by sequence analysis of the PCR products, TaqMan® real-time RT-PCR is a superior method for rapid and simple monitoring of environmental waters for NoVs.

REFERENCES

Anderson, A.D., Garrett, V.D., Sobel, J., Monroe, S.S., Fankhauser, R.L., Schwab, K.J., Bresee, J.S., Mead, P.S., Higgins, C., Campana, J., and R.I. Glass. 2001. Multistate outbreak of Norwalk-like virus gastroenteritis associated with a common caterer. Am. J. Epidemiol. 154(11):1013-1019.

Ando, T., Mulders, M., Lewis, D., Estes, M., Monroe, S., and R. Glass. 1994. Comparison of the polymerase region of small round structured virus strains previously classified in three antigenic types by solid-phase immune electron microscopy. Archives of Virology. 135(1-2):217-226.

Ando, T., S. Monroe, J. Gentsch, Q. Jin, D. Lewis, and R. Glass. 1995. Detection and differentiation of antigenically distinct small round-structured viruses (Norwalk-Like Viruses) by Reverse Transcription-PCR and Southern Hybridization. Journal of Clinical Micrbiology. 33(1): 64–71.

Ando, T., Noel, J., and R. Fankhauser. 2000. Genetic Classification of "Norwalk-like Viruses." The Journal of Infectious Diseases. 181(Suppl 2):S336–48.

Atmar, R.L., Neill, F.H., Romalde, J.L., Le Guyader, F., Woodley, C.M., Metcalf, T.G., and M.K. Estes. 1995. Detection of Norwalk virus and hepatitis A virus in shellfish tissues with the PCR. Appl Environ Microbiol. 61(8):3014-3018.

Atmar, R.L., and M.K. Estes. 2001. Diagnosis of Noncultivatable Gastroenteritis Viruses, the Human Caliciviruses. Clin. Microbiol. Rev. 14 (1):15-37.

Belanger, S.D., Boissinot, M., Menard, C., Picard, F.J., and Bergeron, M.G. 2002. Rapid detection of Shiga toxin-producing bacteria in feces by multiplex PCR with molecular beacons on the smart cycler. J Clin Microbiol 40(4):1436-1440.

Beller, M., Ellis, A., Lee, S.H., Drebot, M.A., Jenkerson, S.A., Funk, E., Sobsey, M.D., Simmons, O.D. 3rd, Monroe, S.S., Ando, T., Noel, J., Petric, M., Middaugh, J.P., and J.S. Spika. 1997. Outbreak of viral gastroenteritis due to a contaminated well. International consequences. JAMA 278(7):563-568.

Bennett, A.R., MacPhee, S., and R.P. Betts. 1996. The isolation and detection of *Escherichia coli* O157 by use of immunomagnetic separation and immunoassay procedures. Lett Appl Microbiol. 22:237-243

Berg, D.E., Kohn, M.A., Farley, T.A., and L.M. McFarland. 2000. Multi-state outbreaks of acute gastroenteritis traced to fecal-contaminated oysters harvested in Louisiana. J. Infect. Dis. 181 Suppl 2:S381-6.

Bertolotti-Ciarlet, A., Crawford, S., Hutson, A., and M. Estes. 2003. The 3' end of Norwalk virus mRNA contains determinants that regulate the expression and stability of the viral capsid protein VP1: a novel function for the VP2 protein. Journal of Virology. 77(21):11603-11615.

Beuret, C. 2004. Simultaneous detection of enteric viruses by multiplex real-time RT-PCR. Journal of Virological Methods. 115:1–8.

Bhattacharya, S.S., M. Kulka, K.A. Lampel, T.A. Cebula, and B.B. Goswami. 2004. Use of reverse transcription and PCR to discriminate between infectious and non-infectious hepatitis A virus. J Virol Methods. 116(2):181-7.

Boccia, D., Tozzi, A.E., Cotter, B., Rizzo, C., Russo, T., Buttinelli, G., Caprioli, A., Marziano, M.L., and Ruggeri, F.M. 2002. Waterborne outbreak of Norwalk-like virus gastroenteritis at a tourist resort, Italy. Emerg. Infect. Dis. 8(6):563-568.

Boom, R., Sol, C., Salimans, M., Jansen, C., Wertheim van Dillen, P., and J. van der Noordaa. 1990. Rapid and Simple Method for Purification of Nucleic Acids. Journal of Clinical Microbiology. 28 (3):495-503.

Brandt, C. D., Kim, H. W., Rodriguez, W. J., Thomas, L., Yolken, R. H., Arrobio, J. O., Kapikian, A. Z., Parrott, R. H., and R. M. Chanock. 1981. Comparison of direct electron microscopy, immune electron microscopy, and rotavirus enzyme-linked immunosorbent assay for detection of gastroenteritis viruses in children. Journal of Clinical Microbiology. 13(5):976-981.

Burton-MacLeod, J., Kane, E., Beard, R., Hadley, L., Glass, R., and T. Ando. 2004. Evaluation and comparison of two commercial enzyme-linked immunosorbent assay kits for detection of antigenically diverse human noroviruses in stool samples. Journal of Clinical Microbiology. 42(6):2587-2595.

Butt, A.A., Aldridge, K.E., and Sanders, C.V. 2004. Infections related to the ingestion of seafood Part I: Viral and bacterial infections. Lancet Infect Dis. 2004 Apr;4(4):201-12. Review.

Cao, S., Lou, Z., Tan, M., Chen, Y., Liu, Y., Zhang, Z., Zhang, X.C., Jiang, X., Li, X., and Z. Rao. 2007. Structural basis for the recognition of blood group trisaccharides by norovirus. J. Virol. Mar 28; [Epub ahead of print].

Croci, L., Losio, M.N., Suffredini, E., Pavoni, E., Di Pasquale, S., Fallacara F., and G. Arcangeli. 2007. Assessment of human enteric viruses in shellfish from the northern Adriatic sea. Int J Food Microbiol. 114(2):252-7.

Cunney, R. J., Costigan, P., McNamara, E. B., Hayes, B., Creamer, E., LaFoy, M., Al Ansari, N., and E. G. Smyth. 2000. Investigation of an outbreak of gastroenteritis caused by Norwalk-like virus, using solid phase immune electron microscopy. Journal of Hospital Infection. 44:113-118.

DeLeon, R., Matsui, S.M., Baric, R.S., Herrmann, J.E., Blacklow, N.R., Greenberg, H.B, and M.D. Sobsey. 1992. Detection of Norwalk virus in stool specimens by reverse transcriptase-polymerase chain reaction and nonradioactive oligoprobes. J. Clin. Microbiol. 30(12):3151-7.

De Medici, D., Croci, L., Suffredini, E., and L. Toti. 2004. Reverse transcription-booster PCR for detection of noroviruses in shellfish. Applied and Environmental Microbiology. 70(10): 6329-6332.

Duizer, E., Bijkerk, P., Rockx, B., de Groot, A., Twisk, F., and M. Koopmans. 2004. Inactivation of Caliciviruses. Applied and Environmental Microbiology. 70(8):4538-4543.

Fankhauser, R., Noel, J., Monroe, S., Ando, T., and Glass. R. 1998. Molecular epidemiology of "Norwalk-like viruses" in outbreaks of gastroenteritis in the United States. Journal of Infectious Diseases. 178(6):1571-1578.

Fankhauser, R., Monroe, S. S., Noel, J. S., Humphrey, C. D., Bresee, J. S., Parashar, U. D., Ando, T., and R. I. Glass. 2002. Epidemiologic and molecular trends of “norwalk-like viruses” associated with outbreaks of gastroenteritis in the United States. The Journal of Infectious Diseases. 186:1-7.

Gallimore, C., Green, J., Richards, A. F., Cotterill, H., Curry, A., Brown, D.W.G, and J. J. Gray. 2004. Methods for the detection and characterization of noroviruses associated with outbreaks of gastroenteritis: outbreaks occurring in the North-West of England during two norovirus seasons. Journal of Medical Virology. 73:280-288.

Gilpatrick, S.G., Schwab, K.J., Estes, M.K., and R.L. Atmar. 2000. Development of an Immunomagnetic captre reverse transcription-PCR assay for the detection of Norwalk virus. J. Virol. Meth. 90:69-78.

Glass, P. J., White, L., Ball, J. M., Leparc-Goffart, I., Hardy, M. E., and M. K. Estes. 2000. Norwalk virus open reading frame 3 encodes a minor structural protein. Journal of Virology. 74(14):6581-6591.

Glass, R., Noel, J., Ando, T., Fankhauser, R., Belliot, G., Mounts, A., Parashar, U. D., Bresee, J. S., and S. S. Monroe. 2000. The epidemiology of enteric caliciviruses from humans: A reassessment using new diagnostics. The Journal of Infectious Diseases. 181: S254-61.

Gonin, P., Couillard, M., and d'Halewyn M.A. 2000. Genetic diversity and molecular epidemiology of Norwalk-like viruses. J. Infect. Dis.182(3):691-697.

Green, J., Gallimore, C.I., Norcott, J.P., Lewis, D., and Brown, D.W.G. 1995. Broadly reactive reverse transcriptase polymerase chain reaction for the diagnosis of SRSV-associated gastroenteritis. J. Med. Virol. 47:392-398.

Green, J., Wright, P., Gallimore, C., Mitchell, O., Morgan-Capner, P., and D. Brown. 1998. The role of environmental contamination with small round structured viruses in a hospital outbreak investigated by reverse-transcriptase polymerase chain reaction assay. Journal of Hospital Infection. 39(1):39-45.

Green, K.Y. 2007. *Caliciviridae:* The Noroviruses. Fields Virology 5th ed. Lippincott Williams & Wilkins. Phildelphia, PA. pg 949-979.

Green, S., Dingle, K., Lambden, P., Caul, E., Ashley, C., and I, Clark. 1994. Human enteric Caliciviridae: a new prevalent small round-structured virus group defined by RNA-dependent RNA polymerase and capsid diversity. Journal of Gen Virology. 75:1883-1888.

Green, S., Lambden, P., Caul, O., Ashley, C., and I, Clark. 1995. Capsid diversity in small round-structured viruses: molecular characterization of an antigenically distinct human enteric calicivirus. Virus Res. 37(3):271-283.

Griffin, D., Donaldson, K. A., Paul, J. H., and J. B. Rose. 2003. Pathogenic human viruses in coastal waters. Clinical Microbiology Reviews. 16(1):129–143.

Green, J., Henshilwood, K., Gallimore, C.I., Brown, D.W.G., and D.N. Lees. 1998. A nested reverse transcriptase PCR assay for detection of small round-structured viruses in environmentally contaminated molluscan shellfish. Appl. Environ. Microbiol. 64:858-863.

Hale, A.D., Green, J., and D.W.G. Brown. 1996. Comparison of four RNA extraction methods for the detection of small round structured viruses in faecal specimens. Journal of Virological Methods. 57:195-201.

Hale, A., Crawford, S., Ciarlet, M., Green, J., Gallimore. C., Brown, D., Jiang, X., and M. Estes. 1999. Expression and self-assembly of Grimsby virus: antigenic distinction from Norwalk and Mexico viruses. Clinical Diagnostics Laboratory Immunology. 6(1):1420145.

Häfliger, D., Gilgen, M., Lüthy, J., and P. Hübner. 1997. Seminested RT-PCR systems for small round structured viruses and detection of enteric viruses in seafood. International Journal of Food Microbiology. 37:27-36.

Häfliger, D., Hubner, P., and Luthy, J. 2000. Outbreak of viral gastroenteritis due to sewage-contaminated drinking water. Int. J. Food Microbiol. 54(1-2):123-126.

Haramoto, E., Katayama H., and S. Ohgaki. 2004. Detection of noroviruses in tap water in Japan by means of a new method for concentrating enteric viruses in large volumes of freshwater. Applied and Environmental Microbiology. 70(4): 2154–2160.

Harrington, P.R., Lindesmith, L., Yount, B., Moe, C.L., and R.S. Baric. 2002. Binding of Norwalk virus-like particles to ABH histo-blood group antigens is blocked by antisera from infected human volunteers or experimentally vaccinated mice. J. Virol. 76(23):12335-12343.

Harrrington, P.R., Vinje, J., Moe, C.L., and R.S. Baric. 2004. Norovirus capture with histo-blood group antigens reveals novel virus-ligand interactions. J. Virol. 78(6):3035-3045.

Hayashi, Y., Ando, T. Utagawa, E., Sekine, S., Okada, S., Yabuuchi, K., Miki, T., and M. Ohashi. 1989. Western blot (immunoblot) assay of small, round-structured virus associated with an Acute Gastroenteritis Outbreak in Tokyo. Journal of Clinical Microbiology. 27(8):1728-1733.

Hedlund, K.O., Rubillar-Abreu, E., and L. Svensson 2000. Epidemiology of calicivirus infections in Sweden, 1994-1998. J. Infect. Dis. 181(Suppl 2):S275-80.

Henshilwood, K., Green, J., and D.N. Lees. 1998. Monitoring the marine environment for Small Round Structured Viruses (SRSVs): a new approach to combating the transmission of these viruses by molluscan shellfish. Water Science and Technology. 38(12):51-56.

Henshilwood, K., Dore, W.J., Anderson, S., and D.N. Lees. 2003. Investigation of norwalk like virus elimination during depuration using a real time quantitative PCR. p.451-465. A. Villalba, J.L. Romalde, B. Reura & R. Beiras (eds) Molluscan Shellfish Safety. 'Conselleria de Pesca e Asuntos Maritimos da Xunta de Galicia and Intergovernmental Oceanographic Commission of UNESCO'

Hill, V.R., Sobsey, M.D., and M. Wu. 2002. Development of alternative methods fro concentrating Norwalk viruses from water. 102nd Annual meeting ASM Salt Lake City May 2002. Q-247.

Hoebe, C.J., Vennema, H., de Roda Husman, A.M., and Y.T. van Duynhoven. 2004. Norovirus outbreak among primary schoolchildren who had played in a recreational water fountain. J Infect Dis. 189(4);699-705.

Hőhne, M., and E. Schreier. 2004. Detection and characterization of norovirus outbreaks in Germany: application of a one-tube RT-PCR using a fluorogenic real-time detection system. J. Med. Virol. 72(2):312-9.

Huang, P., Farkas, T., Marionneau, S., Zhong, W., Ruven-Clouet, N, Morrow, A.L., Altaye, M., Pickering, L.K., Newburg, D.S., Le Pendu, J., and X. Jiang. 2003. Noroviruses bind to human ABO, Lewis and secretor histo-blood group antigens: identification of 4 strain-specific patterns. J. Infect. Dis. 188(1):19-31.

Hutson, A.M., Atmar, R.L., Graham, D.Y., and M.K. Estes. 2002. Norwalk virus infection and disease is associated with ABO histo-blood group type. J. Infect. Dis. 185(9):1335-7.

Hutson, A.M., R.L. Atmar, D.M Marcus, and M.K. Estes. 2003. Norwalk virus-like particle hemagglutination by binding to h histo-blood group antigens. J. Virol. 77(1):405-15.

Jansen, R., G. Siegl, and S. Lemon. 1990. Molecular epidemiology of human hepatitis A virus defined by an antigen-capture polymerase chain reaction method. Proclamation of the National Academy of Science U.S.A. 87(8):2867-2871.

Jiang, X., Graham, D.Y., Wang, K., and M.K. Estes. 1990. Norwalk virus genome cloning and characterization. Science 250:1580-1583.

Jiang, X., Wang, J., Graham, D.Y., and M.K. Estes. 1992. Detection of Norwalk virus in stool by polymerase chain reaction. J. Clin. Microbiol. 30:2529-2534.

Jiang, X., Huang, P.W., Zhong, W.M., Farkas, T., Cubitt, D.W. and D.O. Matson. 1999. Design and evaluation of a primer pair that detects both Norwalk- ans Sapporo-like caliciviruses by RT-PCR. J. Virol. Methods. 83:145-154.

Jiang, X., Huang, P.W., Laborde, D., and D.O. Matson. 2001. Detection of caliciviruses in water samples. AWWA Research Foundation

Jothikumar, N., Cliver, D.O., and D.W. Mariam. 1998. Immunomagnetic capture PCR for rapid concentration and detection of hepatitis A virus from environmental samples Appl. Environ. Microbiol. 64:504-508.

Jothikumar, N., Lowther, J.A., Henshilwood, K., Lees, D.N., Hill, V.R., and J. Vinjé. 2005. Rapid and sensitive detection of noroviruses by using TaqMan-based one-step reverse transcription-PCR assays and application to naturally contaminated shellfish samples. Appl. Environ. Microbiol. 71(4):1870-5.

Kageyama, T., Kojima, S., Shinohara, M., Uchida, K., Fukushi, S., Hoshino, F.B., Takeda, N., and K. Katayama. 2003. Broadly reactive and highly sensitive assay for Norwalk-like viruses based on real-time quantitative reverse transcription-PCR. J. Clin. Microbiol. 41(4):1548-57.

Kageyama, T., Shinohara, M., Uchida, K., Fukushi, S., Hoshino, F.B., Kojima, S., Takai, R., Oka, T., Takeda, N., and K. Katayama. 2004. Coexistence of multiple genotypes, including newly identified genotypes, in outbreaks of gastroenteritis due to Norovirus in Japan. J. Clin. Microbiol. 42(7):2988-95.

Katayama, K., Shirato-Horikoshi H., Kojima, S., Kageyama, T., Oka, T., Hoshino, F., Fukushi, S., Shinohara, M., Uchida, K., Suzuki, Y., Gojobori, T.,and N. Takeda. 2002. Phylogenetic analysis of the complete genome of 18 Norwalk-like viruses. Virology 299(2):225-239.

Kapikian, A., James, H., Kelly, S., King, L., Vaughn, A., and R. Chanock. 1972. Hemadsorption by coronavirus strain OC43. Proclamation Society of Experimental Biological Medicine. 139(1):179-186.

Kapikian, A. 2000. The discovery of the 27-nm norwalk virus: An historic perspective. The Journal of Infectious Diseases. 181(Suppl 2):S295-302.

Karim, M. R., Pontius, F. W., and M. W. LeChevallier. 2004. Detection of noroviruses in water – Summary of an international workshop. Journal of Infectious Diseases. 189:21-28.

Karst, S.M., Wobus, C.E., Lay, M., Davidson, J., and H.W. Virgin 4th. 2003. STAT1-dependent innate immunity to a Norwalk-like virus. Science. Mar 7;299(5612):1575-8.

Khan, A.S., Moe, C.L. Glass, R.I. Monroe, S.S., Estes, M.K., Chapman, L.E. , Jiang, X. , Humphrey, C., Pon, E., Iskander, J.K., et al. 1994. Norwalk virus-associated gastroenteritis traced to ice comsumption aboard a cruise ship in Hawaii: comparison and application of molecular method-based assays. J. Clin. Microbiol. 32(2):318-322.

Kingsley, D., and G. Richards. 2001. Rapid and efficient extraction method for reverse transcription-PCR detection of hepatitis A and Norwalk-like viruses in shellfish. Applied and Environmental Microbiology. 67(9):4152-4157.

Kobayashi, S., Natori, K., Takeda, N., and K. Sakae. 2004. Immunomagnetic capture RT-PCR for detection of norovirus from foods implicated in a foodborne outbreak. Microbiol. Immunol. 48(3):201-204.

Kohn, M.A., Farley, T.A., Ando, T., Curtis, M., Wilson, S.A., Jin, Q.,. Monroe, S.S., Baron, R.C., McFarland, L.M., and R.I. Glass. 1995. An outbreak of Norwalk virus gastroenteritis associated with eating raw oysters. Implications for maintaining safe oyster beds. JAMA. 273(6):466-71.

Kojima, S., Kaheyama, T., Fukushi, S., Hoshino, F.B., Shinohara, M., Uchida, K., Natori, K., Takeda, N., and K. Katayama. 2002. Genogroup-specific PCR primers for detection of Norwalk-like viruses. J. Vir. Methods 100:107-114.

Koopmans, M.P.G., von Bonsdorff, C-H., Vinjé, J., DeMedici, D., and S.S. Monroe. 2002. Foodborne Enteric Viruses. FEMS Microbiology Reviews. 26(2): 187-205.

Koopmans, M., and E. Duizer. 2004. Foodborne viruses: an emerging problem. Int. J. Food. Microbiol. 90(1):23-41.

Kukkula, M., Maunula, L., Silvennoinen, E., and C.H. von Bonsdorff. 1999. Outbreak of viral gastroenteritis due to drinking water contaminated by Norwalk-like viruses. J. Infect. Dis.180(6):1771-1776.

Lambden, P.R,, Caul, E.O., Ashley, C.R., and I.N. Clarke. 1993. Sequence and genome organization of a human Small Round-Structured (Norwalk-like) virus. Science 259:516-519.

Lawson, H.W., Braun, M.M, Glass, R.I., Stine, S.E., Monroe, S.S., Atrash, H.K., Lee, L.E., and S.J. Englender. 1991. Waterborne outbreak of Norwalk virus gastroenteritis at a southwest US resort: role of geological formations in contamination of well water. Lancet 337(8751):1200-4.

Le Guyader, F., Estes, M.K., Hardy, M.E., Neill, F.H., Green, J., Brown, D.W.G., and R.L. Atmar. 1996. Evaluation of a degenerate primer for the PCR detection of human caliciviruses. Arch Virol 141:2225-2235.

Le Guyader, F., Neill, F. Estes, M., Monroe, S., Ando, T., and R. Atmar. 1996. Detection and analysis of a small round-structured virus strain in oysters implicated in an outbreak of acute gastroenteritis. Applied and Environmental Microbiology.62(11):4268-4272.

Le Guyader, F.S., Neill, F.H., Dubois, E., Bon, F., Loisy, F., Kohli, E., Pommepuy, M., and R.L. Atmar. 2003. A semiquantitative approach to estimate Norwalk-like virus contamination of oysters implicated in an outbreak. Int. J. Food. Microbiol. 87(1-2):107-12.

Lodder, W.J. and A. M. de Roda Husman. 2005. Presence of noroviruses and other enteric viruses in sewage and surface waters in The Netherlands. Applied and Environmental Microbiology. 71(3):1453-1461.

Lamothe, G.T., Putallaz, T., Joosten, H., and J.D. Marugg. 2003. Reverse transcription-PCR analysis of bottled and natural mineral waters for the presence of noroviruses. Appl. Environ. Microbiol. 69(11):6541-9.

Laverick, M.A., Wyn-Jones, A.P., and M.J. Carter. 2004. Quantitative RT-PCR for the enumeration of noroviruses (norwalk-like viruses) in water and sewage. Letters in Applied Microbiology. 39: 127–136.

Lees, D.N., Henshilwood, K., and W.J. Dore. 1994. Development of a method for detection of enteroviruses in shellfish by PCR with poliovirus as a model. Appl. Environ. Microbiol. 60(8):2999-3005.

Lees, D.N., Henshilwood, K., Green, J., Gallimore, C.I., and D.W. Brown. 1995. Detection of small round structured viruses in shellfish by reverse transcription-PCR. Appl. Environ. Microbiol. 61(12):4418-24.

Lees, D. 2000. Viruses and bivalve shellfish. Int. J. Food. Microbiol. 59(1-2):81-116.

Lindesmith, L., Moe, C. , Marionneau, S. , Ruvoen, N., Jiang, X., Lindblad, L., Stewart, P., Le Pendu, J., and R. Baric. 2003. Human susceptibility and resistance to Norwalk virus infection. Nat. Med. 9(5):548-553.

Lindesmith, L., Moe, C., LePendu, J., Frelinger, J.A., Treanor, J., and R.S. Baric. 2005. Cellular and humoral immunity following Snow Mountain virus challenge. 79:2900-2909.

Liu, B.L., Clarke, I.N., Caul, E.O., P.R. Lambden. 1995. Human enteric caliciviruses have a unique genome structure and are distinct from Norwalk-like viruses. Arch Virol 140:1345-1356.

Loisy, F., Atmar, R.L., Guillon, P., Le Cann, P., Pommepuy, M. and F.S. Le Guyader. 2005. Real-Time RT-PCR for Norovirus screening in shellfish. J. Vir Methods. J Virol Methods. Jan;123(1):1-7.

Mackay, I.M., Arden, K.E. and A. Nitsche. 2002. Real-time PCR in Virology Nucleic Acids Research 30(6):1292-1305.

Maguire, A.J., Green, J., Brown, D.W.G., Desselberger, U., and J.J. Gray. 1999. Molecular epidemiology of outbreaks of gastroenteritis associated with small round-structured viruses in East Anglia, United Kingdom, during the 1996-1997 season. J. Clin. Microbiol. 37(1):81-89.

Marionneau, S., Ruvoen, N., Le Moullac-Vaidye, B., Clement, M., Cailleau-Thomas, A., Ruiz-Palacois, G., Huang, P., Jiang, X., and J. Le Pendu 2002. Norwalk virus binds to histo-blood group antigens present on gastroduodenal epithelial cells of secretor individuals. Gastroenterology 122(7):1967-1977.

Maunula, L., Kalso, S., Von Bonsdorff, C.H., and A. Ponka. 2004. Wading pool water contaminated with both norovirues and astrovirueses. Epidemiol Infect. 132(4):737-43.

Maunula, L., and C.H. Von Bonsdorff. 2005. Norovirus genotypes causing gastroenteritis outbreaks in Finland 1998-2002. Journal of Clinical Virology. Nov;34(3):186-94.

McDonnell, S., Kirkland, K.B., Hlady, W.G., Aristeguieta, C., Hopkins, R.S., Monroe, S.S., and R.I. Glass. 1997. Failure of cooking to prevent shellfish-associated viral gastroenteritis. Arch. Intern. Med. 157(1):111-6.

Mead, P., Slutsker, L., Dietz, V., McCaig, L., Bresee, S., Shapiro, C., Griffin, P., and R. Tauxe. 1999. Food-related illness and death in the United States. Emerging Infectious Diseases. 5(5):607-625.

Moe, C., Gentsch, J. Ando, T. Grohmann, G., Monroe, S., Jiang, X., Wang, J. Estes, M., Seto, Y. and Humphrey. C. 1994. Application of PCR to detect Norwalk virus in fecal specimens from outbreaks of gastroenteritis. Journal of Clinical Mirobiology. 32(3):642-648.

Moe, C., Sobsey, M. D., Stewart, P., and D. Crawford-Brown. 1999. Estimating the risk of human calicivirus infection from drinking water, p. 4-6, Workshop on Human Calicivirus, Atlanta, Georgia

Moe, C.L. 2001. Effects Assessment: Determinants of Risk and Outcome of Waterborne Infections, p. 115-135. *In* Microbial Pathogens and Disinfection By-products in Drinking Water: Health Effects and Management of Risks. G.F. Craun, F.S. Hauchman and D.E. Robinson (ed.), ILSI Press, Washington, DC.

Mounts, A.W., Ando, T., Koopmans, M., Bresee, J.S., Noel, J., and R.I. Glass. 2000. Cold weather seasonality of gastroenteritis associated with Norwalk-like viruses. J. Infect. Dis.181(Suppl 2):S284-287.

Myrmel, M., Rimstad, E. and Y. Wasteson. 2000. Immunomagnetic separation of a Norwalk-like virus (genogroup I) in artificially contaminated environmental water samples International J. of Food Microbiology 62:17-26.

Myrmel, M., Berg, E.M., Rimstad, E., and B. Grinde. 2004. Detection of enteric viruses in shellfish from the Norwegian coast. Appl. Environ. Microbiol. 70(5):2678-84.

Nakata, S., Honma, S., Numata, K.K., Kogawa, K., Ukae, S., Morita, Y., Adachi, N., and S. Chiba. 2000. Members of the family caliciviridae (Norwalk virus and Sapporo virus) are the most prevalent cause of gastroenteritis outbreaks among infants in Japan. J. Infect. Dis. 181(6):2028-2032.

Nishida, T., Kimura, H., Saitoh, M., Shinohara, M., Kato, M., Fukuda, S., Munemura, T., Mikami, T., Kawamoto, A., Akiyama, M., Kato, Y., Nishi, K., Kozawa, K., and O. Nishio. 2003. Detection, quantitation, and phylogenetic analysis of noroviruses in Japanese oysters. Appl. Environ. Microbiol. 69(10):5782-6.

Noel, J.S., Ando, T., Leite, J.P., Green, K.Y., Dingle, K.E., Estes, M.K., Seto, Y. Monroe, S.S., and R.I. Glass. 1997. Correlation of patient immune responses with genetically characterized Small Round-Structured Viruses involved in outbreaks of nonbacterial acute gastroenteritis in the United States, 1990-1995. J. Med. Virol. 53(4):372-383.

Pang, X.L., Joensuu, J., and T. Vesikari. 1999. Human calicivirus-associated sporadic gastroenteritis in Finnish children less than two years of age followed prospectively during a rotavirus vaccine trial. Pediatr Infect Dis J 18(5):420-426.

Pang, X., Lee, B., Chui, L., Preiksaitis, J.K., and S.S. Monroe. 2004. Evaluation and validation of real-time reverse transcription-pcr assay using the LightCycler system for detection and quantitation of norovirus. J Clin Microbiol. 42(10):4679-85.

Ponka, A., Maunula, L., von Bonsdorff, C.H., and O. Lyytikainen. 1999. An outbreak of calicivirus associated with consumption of frozen raspberries. Epidemiol. Infect. 123(3):469-474.

Parashar, U.D., and S.S. Monroe. 2001. "Norwalk-like viruses" as a cause of foodborne disease outbreaks. Rev. Med. Virol. 11(4):243-52.

Pastorek, J. and K. Marcinka. 1989. Effects of pH and ionic strength on precipitation of phyopathogenic viruses by polyethylene glycol. Acta. Virology. 33(1):68-74.

Reynolds, K.A., Gerba, C.P., and I.L. Pepper. 1995. Detection of enteroviruses in marine waters by direct RT-PCR and cell culture. Water Science Technology. 31(5-6): 323-328.

Richards, G., Watson, M., Fankhauser, R., and S. Monroe. 2004. Genogroup I and II Noroviruses Detected in Stool Samples by Real-Time Reverse Transcription-PCR Using Highly Degenerate Universal Primers. Applied and Environmental Microbiology. 70(12):7179-7184.

Ruvoen-Clouet, N., Ganiere, J.P., Andre-Fontaine, G., Blanchard, D., and J. Le Pendu. 2000. Binding of rabbit hemorrhagic disease virus to antigens of the ABH histo-blood group family. J. Virol. 74(24):11950-11954.

Schvoerer, E., Bonnet, F., Dubois, E., Rogues, A.M., Gachie, J.P., Lafon, M.E., and H.J. Fleury. 1999. A hospital outbreak of gastroenteritis possibly related to the contamination of tap water by a small round structured virus. J. Hosp. Infect. 43(2):149-154.

Schwab, K.J., De Leon, R., and M.D. Sobsey. 1996. Immunoaffinity concentration and purification of waterborne enteric viruses for detection by reverse transcriptase PCR. Appl. Env. Microb. 62(6):2086-2094.

Schwab, K., Estes, M., Neill, F., and R. Atmar. 1997. Use of heat release and an internal RNA standard control in reverse transcription-PCR detection of Norwalk virus from stool samples. Journal of Clinical Microbiology. 35(2):511-514.

Schutz, E., and N. von Ahsen. 1999. Spreadsheet software for thermodynamic melting point prediction of oligonucleotide hybridization with and without mismatches. Biotechniques 27(6):1218-1224.

Shieh, Y.S., Baric, R.S., and M.D. Sobsey. 1997. Detection of low levels of enteric viruses in metropolitan and airplane sewage. Appl. Environ. Microbiol. 63(11):4401-4407.

Shieh, Y-S.C., Monroe, S.S., Fankhauser, R.L., Langlois, G.W., Burkhardt III, W., and R.S. Baric. 2000. Detection of Norwalk-like virus in shellfish implicated in illness. J. Infect. Dis. 181(Suppl 2):S360-366.

Shieh, Y.C., Baric, R.S., Woods, J.W, and K.R. Calci. 2003. Molecular surveillance of enterovirus and Norwalk-like virus in oysters relocated to a municipal-sewage-impacted gulf estuary. Appl Environ Microbiol. 69(12):7130-6.

Sobsey, M.D. and J.S. Glass. 1980. Poliovirus concentration from tap water with electropositive adsorbent filters. Appl. Environ. Microbiol. 40(2):201-210.

Sobsey, M.D. and A.R. Hickey. 1985. Effects of humic and fulvic acids on poliovirus concentration from water by microporous filtration. Appl. Environ. Microbiol. 49(2):259-264.

Straub, T.M., K.H. zu Bentrup, K.H., Orosz-Coghlan, P., Dohnalkova, A., Mayer, B.K., Bartholomew, R.A., Valdez, C.O., Bruckner-Lea, C.J., Gerba, C.P., Abbaszadegan, M., and C.A. Nickerson. 2007. In vitro cell culture infectivity assay for human noroviruses. Emerg. Infect. Dis. 13(3):396-403.

Sunen, E., and M. Sobsey. 1999. Recovery and detection of enterovirus, hepatitis A virus and Norwalk virus in hardshell clams (Mercenaria mercenaria) by RT-PCR methods. Journal of Virological Methods. 77(2):179-187.

Tan, M., P. Huang, P., Meller, J., Zhong, W., Farkas, T. and X. Jiang. 2003. Mutations within the P2 domain of Norvirus capsid affect binding to human histo-blood group antigens: evidence for a binding pocket. J. Virol. 77(23):12562-12571.

Tan, M., Hegde, R.S., and X. Jiang. 2004. The P domain of Norovirus capsid protein forms dimer and binds to histo-blood group antigen receptors. J. Virol. 78(12):6233-6242.

Trujillo AA, McCaustland KA, Zheng DP, Hadley LA, Vaughn G, Adams SM, Ando T, Glass RI, Monroe SS. 2006 Use of TaqMan real-time reverse transcription-PCR for rapid detection, quantification, and typing of norovirus. J Clin Microbiol. 2006 Apr;44(4):1405-12.

Ueki, Y., Sanoa, D., Watanabea, T., Akiyamab, K. and T. Omuraa. 2005. Norovirus pathway in water environment estimated by genetic analysis of strains from patients of gastroenteritis, sewage, treated wastewater, river water and oysters. Water Research. 39: 4271–4280.

U. S. Environmental Protection Agency. 1999. Method 1623: *Cryptosporidium* and *Giardia* in water by filtration/IMS/FA. (EPA 821-R-99-006) Office of Water, U.S. Environmental protection Agency, Washington, D.C.

Van den Berg, H., Lodder, W., van der Poel, W., Vennema, H. and A. M. de Roda Husman. 2005. Genetic diversity of noroviruses in raw and treated sewage water. Researc ih in Microbiology. 156:532–540.

Vinjé, J., and M.P.G. Koopmans. 1996. Molecular detection and epidemiology of small round-structured viruses in outbreaks of gastroenteritis in The Netherlands. J. Infect. Dis. 174 (3): 610-615.

Vinje, J., Altena, S., and M. Koopmans. 1997. The incidence and genetic variability of small round-structured viruses in outbreaks of gastroenteritis in The Netherlands. Journal of Infectious Diseases. 176(5):1374-1378.

Vinjé, J., and M.P.G. Koopmans. 2000. Simultaneous detection and genotyping of Norwalk-like viruses by oligonucleotide array in a reverse line blot hybridization format. J. Clin. Microbiol. 38(7):2595-2601.

Vinje, J., J. Green, J., Lewis, D.C., Gallimore, C.I., Brown, D.W.G., and M.P.G. Koopmans. 2000. Genetic polymorphism across regions of the three open reading frames of "Norwalk-like viruses." Archives of Virology. 145: 223–241.

Vinjé, J., H. Vennema, L. Maunula, C-H. von Bonsdorff, M. Hoehne, E. Schreier, A. Richards, J. Green, D. Brown, S.S. Beard, S.S. Monroe, E. de Bruin, L. Svensson, and M.P. Koopmans. 2003. International collaborative study to compare reverse transcriptase PCR assays for detection and genotyping of noroviruses. J. Clin. Microbiol. 41(4):1423-33.

Vinjé, J., Hamidjaja, R.A., and M. D. Sobsey. 2004. Development and application of a capsid VP1 (region D) based reverse transcription PCR assay for genotyping of genogroup I and II noroviruses. Journal of Virological Methods. 116: 109–117.

De Wit, M. A. S., Koopmans, M. P. G. , Kortbeek, L. M. , Wannet, W. J. B., Vinjé, J., van Leusden, F., Bartelds, A.I.M., and Y. T. H. P. van Duynhoven. 2001. Sensor, a population-based cohort study on gastroenteritis in the Netherlands: incidence and etiology. American Journal of Epidemiology. 154(7):666-674.

Wang, J., Jiang, X., Madore, H.P., Gray, J., Desselberger, U., Ando, T., Seto, Y., Oishi, I., Lew, J.F., Green, K.Y., and M.K. Estes 1994. Sequence diversity of small, round structured viruses in the Norwalk virus group. J Virol 68:5982-5990.

Widdowson, M.A., Sulka, A., Bulens, S.N., Beard, R.S., Chaves, S.S., Hammond, R., Salehi, E.D., Swanson, E., Totaro, J., Woron, R., Mead, P.S., Bresee, J.S., Monroe, S.S., and R.I. Glass. 2005. Norovirus and foodborne disease, United States, 1991-2000. Emerg Infect Dis. Jan;11(1):95-102.

Wit de, M.A.S., Koopmans, M.P.G., Kortbeek, L.M., Wannet, W.J.B., Vinjé, J., van Leusden, F., Bartelds, A.I.M., and Y.T.H.P. van Duijnhoven. 2001. Sensor, a population-based cohort study on gastroenteritis in the Netherlands: Incidence and etiology. Am. J. Epi. 154:666-674.

Wyn-Jones, A.P., Pallin, R., Dedoussis, C., Shore, J., and J. Sellwood. 2000. The detection of small round-structured viruses in water and environmental materials. Journal of Virological Methods. 87: 99–107.

Yamada, O., Matsumoto, T., Nakashima, M., Hagari, S., Kamahora, T., Ueyama, H., Kishi, Y., Uemura, H. and T. Kurimura. 1990. A new method for extracting DNA or RNA for polymerase chain reaction. Journal of Virological Methods. 27(2):203-209.

Zheng, D.P., Ando, T., Fankhauser, R.L., Beard, R.S., Glass, R.I., and S.S. Monroe. 2006. Norovirus classification and proposed strain nomenclature. Virology. Dec 8; Mar 15;346(2):312-23.

IWA Publishing's authorised EU representative for General Product Safety Regulations is Diane D'Arras, 15 rue Duret, 75116 Paris, France, e-mail: safety@iwap.co.uk.

Printed and bound by CPI Group (UK) Ltd, Croydon, CR0 4YY
17/09/2026
02218798-0007